# A Key for Identification of Rock-forming Minerals in Thin-Section

# A Key for Identification of Rock-forming Minerals in Thin-Section

Andrew J. Barker
*University of Southampton, UK*

**CRC Press**
Taylor & Francis Group
Boca Raton London New York

CRC Press is an imprint of the
Taylor & Francis Group, an **informa** business

CRC Press/Balkema is an imprint of the Taylor & Francis Group, an informa business

© 2014 Taylor & Francis Group, London, UK

Typeset by diacriTech Limited, Chennai, India
Printed and bound in Great Britain by Ashford Colour Press Ltd.

Published by: CRC Press/Balkema
            P.O. Box 11320, 2301 EH Leiden, The Netherlands
            e-mail: Pub.NL@taylorandfrancis.com
            www.crcpress.com – www.taylorandfrancis.com

Library of Congress Cataloging-in-Publication Data

Barker, Andrew J., author.
   A key for identification of rock-forming minerals in thin section / Andrew J. Barker, University of Southampton, UK.
      pages cm
   Includes bibliographical references and index.
   ISBN 978-1-138-00114-5 (paperback)—ISBN 978-1-315-77869-3 (ebook)
   1. Rock-forming minerals—Identification. I. Title.

   QE397.B37 2014
   549'.114—dc23

                                            2014024200

ISBN: 978-1-138-00114-5 (Pbk)
ISBN: 978-1-315-77869-3 (eBook PDF)

# Contents

*Acknowledgements*                                                     *vii*

## 1. Introduction                                                        1

1.1   *Photographs and their captions*                                    3
1.2   *Crystal systems*                                                   3
      1.2.1   *Miller indices*                                            4
1.3   *Mineral properties in plane-polarised light (PPL)*                 5
      1.3.1   *Shape and form*                                            5
      1.3.2   *Textural relationships*                                    5
      1.3.3   *Mineral cleavage*                                          7
      1.3.4   *Colour and Pleochroism*                                    9
      1.3.5   *Relief*                                                   11
1.4   *Mineral properties in cross-polarised light (XPL)*               14
      1.4.1   *Birefringence and interference colours*                  14
      1.4.2   *Twinning*                                                 19
      1.4.3   *Zoning*                                                   20
      1.4.4   *Extinction angles*                                        21
      1.4.5   *Length-fast and length-slow crystals*                     24
      1.4.6   *Interference figures*                                     27

## 2. Key to rock-forming minerals in thin-section                       33

2.1   *To decide which Section*                                          33
2.2   *Symbols & abbreviations*                                          34

**Section 1:   2 (or 3) cleavage traces**                                35

**Section 2:   1 Cleavage trace, inclined extinction**                   53

**Section 3:   1 Cleavage trace, straight extinction, colourless**       71

**Section 4:   1 Cleavage trace, straight extinction, coloured**         87

**Section 5:   Imperfect cleavage, parting or arranged fractures**       97

Section 6:   0 Cleavage traces, colourless                    111

Section 7:   0 Cleavage traces, coloured                      131

Section 8:   Opaque minerals                                  145

### Appendices:

*Appendix 1: Birefringence value (δ) and corresponding interference colours for a 30µm rock thin-section.*                    153
*Appendix 2: Birefringence table of minerals in order of δ values.*    154
*Appendix 3: Glossary.*                                       158
*Appendix 4: Mineral abbreviations.*                          167
*References*                                                  169
*Index*                                                       171

# Acknowledgements

The idea for a key to minerals under the microscope began many years ago, but due to various other commitments it has taken a while to complete. My wife, Linda, has given me encouragement and support throughout, so I would like to start by giving her my sincere thanks. This book would not have been possible without her.

Secondly, I would particularly like to thank all undergraduate and postgraduate students past and present who have taken my mineralogy and petrology courses at the University of Southampton. Their questions, comments and suggestions have guided me greatly in developing the present mineral key. Their interest and enthusiasm has certainly encouraged me throughout my career, and I hope this book lives up to expectations as an aid to mineral identification under the microscope.

I would also like to acknowledge the University of Southampton, for access to its excellent thin-section collection, build up over many years, and from which the vast majority of the minerals photographed were sourced. In this regard, I would also like to give special thanks to Rex Taylor, for the use of his microscope and camera, and for helpful feedback on early drafts of the key. I would also like to thank John Ford and Bob Jones for their support with thin-section preparation over many years.

Special thanks are also extended to colleagues at other Universities who have helped in providing photographs and thin-sections of minerals that I had found difficult to source. In this regard I would like to pay special thanks to Giles Droop (University of Manchester) for his considerable effort in providing many excellent photographs of minerals I was looking for. He also reviewed an early draft of the manuscript, and provided constructive criticism and helpful suggestions for improvements to the key. Others who have helped with thin-section materials and advice include Colin Wilkins and Arjan Dijkstra (University of Plymouth), Dean Bullen (University of Portsmouth) and Dave Waters (University of Oxford).

Finally, I would like to thank Jenny and Michael Manson for their enthusiasm and support of this publication. Needless to say I've tried to ensure complete accuracy throughout, but given the nature of the challenge I accept full responsibility for any errors or omissions that may have occurred.

To all those mentioned, and to anyone I might have inadvertently overlooked, I would like to say a big thank you.

Andrew Barker, June 2014, Southampton

# Chapter 1

# Introduction

Being able to accurately identify minerals in thin-section is an essential skill for any geologist, and is vital for the correct interpretation of rocks and their petrogenesis. It can be a particularly challenging prospect for students new to the subject. However, with a basic understanding of microscopy, petrology and optical mineralogy, coupled with some reference text books, it is soon possible to identify some of the major rock-forming minerals with confidence. However, being able to move beyond this, to efficiently and reliably identify newly encountered unknown minerals can prove difficult. Students may resort to flicking through the pages of one of the many beautifully illustrated atlases of rock-forming minerals (e.g. MacKenzie & Adams, 1994; MacKenzie & Guilford, 1980) or may use an optical mineralogy book such as Kerr (1977), or the superb *"Introduction to Rock-forming Minerals"* by Deer et al. (1992, 2013). To the novice, however, the latter can seem impenetrable at first glance, but with time becomes an essential companion. It is hoped that by producing a key to mineral identification, an efficient and systematic approach to identifying minerals in thin-section will be provided. This is not intended to replace the detailed optical mineralogy text book, but rather to complement it.

More than a century ago, Johannsen (1908) produced a *"Key for the determination of rock-forming minerals in thin sections"*, which although largely forgotten is a very useful publication. Johanssen (1908) used combinations of various optical properties to separate minerals into small groups. Although not providing a unique solution, it gives the reader a short-list of possibilities, that can be further refined with reference to other optical properties not previously considered. In terms of a "modern" equivalent to this, the closest would be the "Mineral Tables" in Chapter 10 of Kerr (1977), but even this is now more than 35 years old. Like Johanssen's key, the tables of Kerr (1977) and Ehlers (1987b) provide the petrologist with a systematic approach to narrow down to a group of minerals with similar properties, after which, the detailed optical characteristics of each should be considered before concluding on the identity of the mineral in question. The present author has always found Kerr (1977) to be an excellent text, and very easy to use, but once again, the Tables of this text do not provide a unique solution for mineral identification.

The present key, comprises eight "Sections", and as far as possible attempts to key out to a specific mineral. The key assumes that the user has a basic understanding of optical mineralogy and is able to determine essential mineral properties in thin-section using plane-polarised and cross-polarised light. Even so, a brief summary of mineral properties in thin-section is provided in the introduction. In some cases it may be necessary to determine an optic figure for closely similar minerals. However, this is something not routinely employed as it is technically more challenging, and may not be feasible in small, deformed, or heavily included crystals. If this point is reached and uncertainty remains concerning the mineral identification, consider all remaining possibilities (usually only two or three minerals). It may be necessary to employ analytical techniques if the identity of the mineral is still in question.

The present key is designed in a manner comparable to dichotomous botanical keys such as Clapham *et al.* (1981) and Stace (1999), which are widely used by botanists in the UK. It involves a stepped sequence defining characteristic features to arrive at a unique mineral identification. At this point, it should be stated that the present author strongly encourages all those using the key to start by building up a full portfolio of properties for a given mineral before trying to identify it. Only then should the stepped sequence of the key be used.

Starting from couplet 1 (in the appropriate Section), successive numbered steps give two options for a particular characteristic (sometimes with letter-coded illustration). Decide which option fits the mineral being identified and go to the next step as indicated by the number to the right-hand side. Repeat the process until the point where a mineral is named. Having keyed-out the mineral, cross-check with an optical mineralogy text book (eg Kerr, 1977; Ehlers, 1987a, b; Deer *et al.*, 1992, 2013), to confirm that all the properties agree. Also check that the rock and mineral assemblage seem realistic. If the result seems impossible or unlikely, back-track to points of uncertainty. For example, if uncertain whether a particular mineral is moderate or high relief, follow what you think first of all (eg high), but if this fails to produce a realistic answer, back track and follow the other route (ie moderate relief). Please also recognise that whilst 150+ of the most common rock-forming minerals are included in the key, other minerals are not, so some minerals may not key-out.

The key has been designed in such a way as to prioritise those properties that are most easily defined, and thus less ambiguous. It is hoped that this will minimise the chance of misidentification and enable the less experienced petrologist to use the key with confidence. Since in many cases (eg small crystals) the idealised crystal form and characters may not be encountered, the key attempts to "error-trap" by including some minerals in more than one Section (eg plagioclase feldspars may seemingly lack cleavage and often not show twinning in metamorphic rocks). Recognising that end-sections and side sections of minerals such as amphiboles and pyroxenes have different properties, they can be keyed out in different Sections. To aid in the identification of closely similar minerals that have very different petrogenesis, the bulk rock chemistry, environment of formation, and P-T-fluid conditions may be referred to. To the purist some aspects of this approach may seem rough and ready, but to the student or practising geologist who does not have the time or suitable crystals to check 2V optic axial angles and optic sign, this guide should prove highly useful. It should be stated, however, that when using shape, textural relationships or occurrence as primary identifiers, caution must be exercised. Such characteristics can be broadly useful, in the same way that in botanical keys plants may have an association with a particular habitat or geographical area. Ultimately, however, it is the anatomical features of the plant and optical properties of the mineral that are critical to the correct identification. In all cases, therefore, once a mineral identification has been arrived at using the key, it is recommended that a final check of all properties is made with reference to an optical mineralogy text such as Kerr (1977) or Deer *et al.* (1992, 2013).

To the beginner, it is hoped that the key provides a systematic approach to defining unknown minerals, rather than simply defining a few of the optical properties then searching through a mineralogy or petrology text hoping to find a mineral that matches. A glossary is given at the back to provide definitions of various technical terms used in the key and introductory sections. Also included are a tabulation of mineral abbreviations used on photographs, a polarisation colour chart, and birefringence values for all minerals referred to in the key.

## 1.1    Photographs and their captions

Unless otherwise stated, all photographs have been taken by the author. In addition to naming the mineral, each image includes details of the image magnification, a scale-bar, and whether taken in plane-polarised light (PPL), cross-polarised light (XPL), or reflected light (RL). Wherever possible, brief details are given of the rock type and sample location.

### *Image magnifications and scale-bars*

The majority of photographs were taken at x50 or x100 magnification (as specified), with the original horizontal fields of view in such images being 2.7 mm and 1.35 mm, respectively. Unless specified otherwise, all scale-bars represent a distance of 0.25 mm on the image. Where higher magnifications (x400, field of view 0.34 mm, i.e. 340 μm) and close-ups are shown, the field of view is considerably smaller and an appropriate scale-bar is given.

## 1.2    Crystal systems

Before proceeding to consider rock-forming minerals in thin-section using a polarising microscope it is useful to firstly consider the fundamental crystals systems and the inherent link with particular crystal shapes, forms and mineral cleavage orientations. It is outside the scope of the present text to cover this in detail, but a basic awareness of angular and length relationships for crystals of particular systems is helpful when identifying minerals based on cleavage orientations, and crystal shapes formed  by thin-sections cut both perpendicular and oblique to crystallographic axes.

The schematic illustrations of Fig. 1.1 show the principal crystal systems of rock-forming minerals. The varying fundamental relationships between principal cell lengths $a$, $b$, $c$, and angles $\alpha$, $\beta$, $\gamma$ are specified for each system beneath the respective diagrams. The principal crystallographic axes $x$, $y$, $z$ are also shown. To avoid any confusion, please note that the present author has used the $x$, $y$, $z$ notation for consistency with Deer *et al.* (1992, 2013), but in other texts (e.g. Kerr, 1977; Ehlers 1987a, b), the principal crystallographic axes are labelled $a$, $b$, $c$.

In the rather special case of the hexagonal system, rather than defining the cell dimensions in terms of two horizontal axes ($a$, $b$) as in other systems, there are three axes ($a_1$, $a_2$, $a_3$) of equal length and angular relationship (120°), perpendicular to the

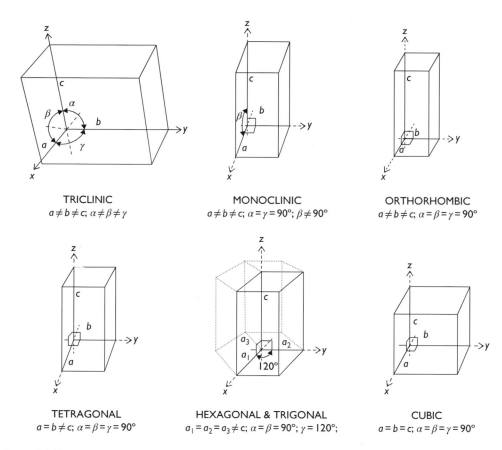

*Figure 1.1* Various crystal systems represented as unit cell with lattice parameters $a$, $b$, $c$ (lengths) and $\alpha$, $\beta$, $\gamma$ (angles), with crystallographic axes $x$, $y$, $z$.

vertical $c$-axis (e.g. Beryl). For further detail on crystal systems of minerals, the text by Wenk & Bulakh (2004) provides a useful introduction.

### 1.2.1    Miller indices

Miller (1839) devised a convenient method for describing the lattice planes of a mineral with reference to the orientation of the plane with respect to the three principal crystallographic axes ($x$, $y$, $z$ of Fig. 1.1). With the exception of the slightly more complicated hexagonal system (see above), which has four axes, and thus four digits, the index is given as a three digit integer value. The figures are derived from the reciprocal values of the axis intercepts, normalised to the lowest common denominator to give integers.

If we consider the cubic system illustrated in Fig. 1.1, the diagram below (Fig. 1.2a) depicts a simple cubic mineral (e.g. pyrite, fluorite) with Miller indices given for the three visible faces. The top face, being parallel with axes $x$ and $y$ and perpendicular to $z$ has the Miller index (001). The face front left is perpendicular to $x$ and parallel to $y$ and $z$, so has the idex (100), and finally the face front right is parallel to $x$ and $z$ and perpendicular to $y$ so has the index (010).

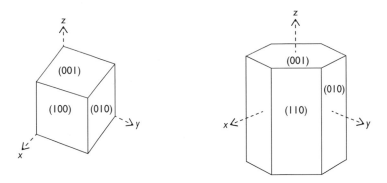

*Figure 1.2* Miller indices for selected faces on a) simple cube (e.g. pyrite, fluorite), and b) staurolite.

For faces perpendicular to the negative end of a particular axis the notation is with a bar above the integer, so the face on the bottom of the cube (Fig. 1.2a) illustrated (not visible), would be (00$\bar{1}$), the one back right (hidden) would be ($\bar{1}$00), and the one back left (hidden) would be (0$\bar{1}$0). A face at a 45° angle to (100) and (010) would have the value (110), (see Fig. 1.2b) and so on. It is outside the scope of present text to provide a more detailed account of Miller indices, but for further information see Wenk and Bulakh (2004). The reason for briefly mentioning Miller indices in this introduction is because on some occasions within the mineral key, reference is made to particular crystal faces, cleavage or twinning in specific planes using Miller indices. For example, see couplet 2. of Section 1.

## 1.3   Mineral properties in plane-polarised light (PPL)

### 1.3.1   Shape and form

Many crystals, especially those of fine grained aggregates, or forming the matrix of rocks, have rather poor crystal form, and would be described as anhedral or irregular. Other crystals, however, may show rather good crystal form, indicative of the crystal system of the particular mineral concerned. This is especially true of phencrysts, porphyroblasts and vein minerals, where unimpeded growth has occurred. If some of the crystal faces are present, the crystal is said to be subhedral, and if all sides of the crystal are well formed it is described as euhedral. Fig. 1.3 illustrates a few examples of distinctive crystal shapes, and the mineral key itself includes schematic illustrations of all shapes and forms as necessary. For more detailed information on the topic see Barker (1998, pp 65–75), MacKenzie *et al.* (1982), and Table 10.3 of Kerr (1977). The latter provides a very useful summary of minerals commonly observed with euhedral form, grouped according to form (e.g. acicular, lath-like, radiating).

### 1.3.2   Textural relationships

As well as general shape providing a clue to the identity of well-formed minerals, there are also textural features that can sometimes further aid identification, and may sometimes be referred to in the mineral key presented in part 2. Some of these features such

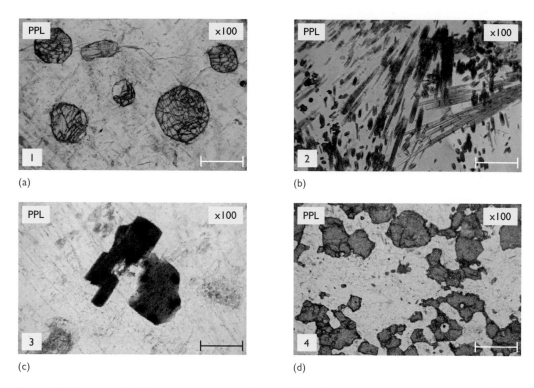

*Figure 1.3* Examples of different crystal shapes/forms: (a) Spodumene end-section; the classic euhedral octagonal shape of pyroxenes. Granite pegmatite; Leinster, Ireland. (b) Radiating acicular tourmaline (schorlite); Luxullyanite, Cornwall, England. (c) Two biotite crystals; the one on the left showing the typical subhedral lath-shaped side-section of biotite with {001} parallel cleavage traces along the length of the crystal, and parallel with crystal sides, whereas the subhedral, roughly hexagonal crystal to the right, with no cleavage, is a view straight down the z-axis (≡ c-axis) of the crystal, thus displaying the (001) face (ie end-section). Granite; St. Jacut, Brittany, France. (d) Anhedral spessartine garnet; Nisserdal, Telemark, Norway.

as porphyroblasts, phenocrysts, and colour zoned minerals due to chemical zonation, are evident in plane-polarised light (PPL). Other features, however, such as twins, and chemical zonation in colourless minerals (e.g. plagioclase, vesuvianite) only become more apparent in cross-polarised light (XPL) due to changes in interference colours (see 1.4.2 and 1.4.3 below). Mineral intergrowths can be seen in PPL if the intergrown phases have contrasting optical properties, but for minerals with similar properties such intergrowths are only obvious in XPL (see myrmekite image 116b below).

Phenocrysts of igneous rocks, and porphyroblasts of metamorphic rocks, are easily observed and, often easily identified, by virtue of their large size providing good information on most if not all of the mineral's optical properties. Their occurrence as phenocrysts or porphyroblasts in itself can be strongly indicative of particular minerals. Table 1 gives a list of commonly observed phenocrysts and porphyroblasts of igneous and metamorphic rocks, and Fig. 1.4 illustrates several representative examples. If the phenocryst or porphyroblast has abundant inclusions, the respective terms poikilitic and poikiloblastic are used (see Fig. 1.4c).

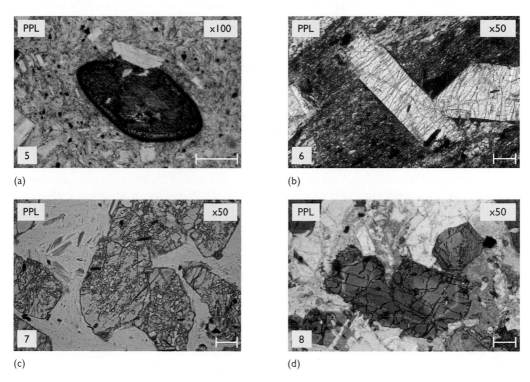

(a)  (b)

(c)  (d)

*Figure 1.4* Textural relationships in plane-polarised light; a) Oxo-hornblende phenocryst in andesite; Mt.Shasta, CA, USA. b) Andalusite (var. chiastolite) porphyroblasts in chiastolite slate; Skiddaw Granite aureole, Lake District, England. c) Poikiloblastic staurolite porphyroblasts in stt-and-schist; Nauyago, Ghana. d) Colour zonation in titanaugite phenocryst in zeolitised basalt; Scawt, Antrim, N.Ireland.

*Table 1.1* Minerals commonly occurring as phenocrysts or porphyroblasts.

| Phenocrysts *(igneous rocks)* | | Porphyroblasts *(metamorphic rocks)* | |
|---|---|---|---|
| Hornblende | Augite | Garnet | Hornblende |
| Biotite | Olivine | Staurolite | Actinolite |
| Quartz | Pigeonite | Andalusite | Lawsonite |
| Plagioclase | Orthopyroxene | Kyanite | Plagioclase (albite) |
| Orthoclase | Anorthoclase | Biotite | K-feldspar |
| Sanidine | Oxo-Hornblende | Chlorite | Pyrite |
| Nepheline | | Chloritoid | Dolomite |
| Leucite | | Cordierite | |
| Analcite | | Muscovite | |

### 1.3.3 Mineral cleavage

The distinction between those minerals that show cleavage from those that do not is one of the most useful characteristics for distinguishing some of the major mineral groups in both hand-specimen and thin-section. Minerals such as amphiboles, pyroxenes and micas have well developed cleavage, whereas common minerals such as quartz, garnet and olivine lack cleavage. In thin-section, the trace of cleavage is observed, and depending on the crystal orientation with respect to the thin-section cut one or more cleavage traces will be seen. Minerals such as micas (e.g. muscovite and biotite) display a single well developed cleavage trace {001} in almost all crystals within a thin-section (Fig. 1.5a). The only exceptions are crystals cut parallel to {001}, which show no cleavage (see hexagonal biotite crystal in Fig. 1.3c). Other minerals such as pyroxenes and amphiboles, have two well developed cleavages, and whilst only displaying a single cleavage trace in side-sections, generally show two well developed cleavages in end-sections (see Fig. 1.5b). The end-section angular relationships between cleavage traces are particularly useful for distinguishing pyroxenes from amphiboles. The former displays cleavage intersecting at approximately 90°, whereas the latter gives an acute angle cleavage intersect of approximately 60°. Presence or absence of cleavage provides such an important division of minerals, and for those minerals with cleavage there is the further possibility of dividing according to the number of cleavages present. In view of this, the number of cleavage traces present, provides much of the basis for construction of the broad sub-divisions ("Sections") of the present mineral key (see 2.1 below). Accordingly, identifying cleavage (or lack of it is critical to deciding which Section of the present Key to start in (see pg.33 below).

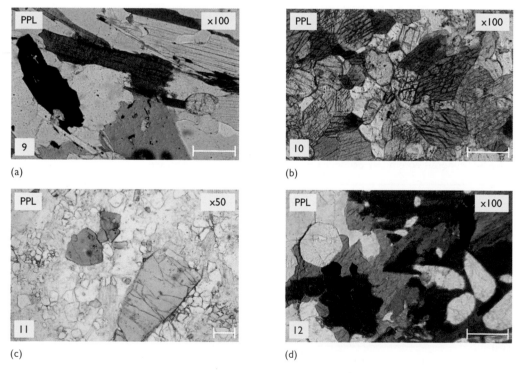

(a)    (b)

(c)    (d)

*Figure 1.5*   Cleavage traces in minerals. a) Single cleavage traces {001} in biotite and muscovite. Biotite schist; Dublin, Ireland. b) Two cleavage traces {110} at 56° and 124° in end-sections of hornblende. Hornblende gneiss; Sleat, Skye, Scotland. c) Irregular but somewhat arranged fractures (not cleavage) in oblique section of tourmaline (main crystal, lower right). Also note the end-section of yellow-green tourmaline (left centre) showing the classic triangular form with curved convex-outward faces. Tourmaline granite; Cornwall, England. d) Colourless apatite crystals with no cleavage, including euhedral hexagonal crystal (left) in larvikite; Dalheim, Larvik, Norway.

### 1.3.4    Colour and Pleochroism

Whether a mineral is coloured or colourless in plane-polarised light is a difference, which forms the basis for several of the broad sub-divisions ("Sections") in the present key for mineral identification. Furthermore, many coloured minerals show pleochroism (i.e. mineral colour changes as the microscope stage is rotated in PPL). Such variations are due to the orientation of the crystal (and its lattice) with respect to the two rays of plane-polarised light.

At this point it is necessary to give a brief explanation of the relationships between optic axes (vibration directions) $\alpha$, $\beta$, $\gamma$ , and crystallographic axes $x$, $y$, $z$, previously discussed (1.2 above). It may seem slightly confusing that $\alpha, \beta, \gamma$ are used for optic axes, as well as the angles between $x$, $y$, $z$ crystallographic axes (see Fig. 1.1). However, the present author has followed the convention of this notation to be consistent with Deer *et al.* (1992, 2013). To avoid any confusion, please note that the published literature is inconsistent on this matter, and sources such as Kerr (1977) and Ehlers (1987a, b) use X, Y, Z for $\alpha, \beta, \gamma$.

Deer *et al.* (1992, 2013) contains extensive information on all rock-forming minerals, and is highly recommended as a final check for mineral identifications arrived at using the present key. Please note that all further reference to $\alpha$, $\beta$, $\gamma$ in the present publication is always in relation to the principal optic axes (vibration directions). The optic axes, which are at 90° to each other, also show a direct relationship with variations in refractive index, with $\alpha$ being the smallest refractive index, $\gamma$ being the largest, and $\beta$ being an intermediate value.

The relationships and terminology described in the preceding paragraph are those for biaxial minerals; namely those belonging to the orthorhombic, monoclinic and triclinic mineral systems. These account for the vast majority of minerals, but for uniaxial minerals (those of tetragonal, hexagonal and trigonal systems), a slightly different arrangement exists. These minerals have a unique orientation, namely the $z$ ($\equiv c$) vertical crystallographic axis, along which the crystal appears isotropic. The axes in the horizontal plane are equal to one another (see Fig. 1.1) and thus have the same refractive index. The index of refraction of the light ray vibrating parallel to the $z$-axis is by convention for uniaxial minerals given the notation $\varepsilon$ (epsilon), whereas the vibration direction perpendicular to this in the horizontal plane has the symbol $\omega$ (omega) (see Ehlers, 1987a for further detail). For cubic minerals, the lattice dimensions and atomic arrangements are identical in all directions, giving three crystallographic axes perpendicular to one another and of equal dimension. This in turn gives a constancy of refractive indices, which in turn gives optic axes that are identical and parallel with crystallographic axes.

Returning to consider pleochroic schemes of minerals, there are specific colours associated with particular orientations of a given mineral. Because of the relationships outlined in the preceding paragraphs, coloured minerals of the cubic system have identical refractive indices in all directions, so do not show pleochroism. Coloured uniaxial minerals (e.g. tourmaline) may show two different colours in a pleochroic scheme by virtue of the difference between the two optic axes $\varepsilon$ and $\omega$, whereas biaxial minerals (e.g. hornblende) may show up to three different colours as they have three different axes $\alpha, \beta, \gamma$.

Pleochroic colours in relation to specific optic axes may be relatively easily determined for euhedral and subhedral elongated crystals with distinctive length and end-sections. Where minerals are anhedral and granular, although it is not usually possible to determine the specific orientation of individual crystals, it should usually be possible to deduce the pleochroic scheme by rotating the microscope stage and looking for different colours in the same field of view (see Fig. 1.6d).

Fig. 1.6 illustrates a few examples of pleochroic minerals. In Fig. 1.6a, b the mineral illustrated is biotite, a common mineral that soon becomes familiar to all petrologists. With the polariser oriented E-W, the biotite (similarly aligned) shows a strong red-brown colour, whereas by rotating the stage, and thus minerals, through 90° (polariser maintained with E-W orientation), the biotite colour changes to pale brown or yellow-brown (Fig. 1.6b). This is a consistent relationship, so if you are not sure which way the polariser is oriented for a given microscope, take a known biotite-bearing sample and check which orientation of biotite produces the strongest colour. Once done, the pleochroic schemes of other unknown minerals can be determined in relation to particular optic axes, such as actinolite (Fig. 1.6c), and piemontite (Fig. 1.6d).

(a)

(b)

(c)

(d)

Figure 1.6 Colours and pleochroism. a) Biotite in garnet mica schist; Ross of Mull, Scotland. b) The same biotite as a), but rotated through 90° to illustrate pleochroism. The double-headed arrows beside the images indicate the polarisation vibration directions. c) Actinolite crystals in a range of orientations, showing the pleochroic scheme $\alpha$ = yellow-green, $\beta$ = green, $\gamma$ = dark green/blue-green; Epidote-actinolite hornfels; locality unknown. d) Piemontite, showing a spectacular pleochroic scheme: $\alpha$ = orange, $\beta$ = violet, $\gamma$ = red; Pennsylvania, USA.

### 1.3.5    Relief

An accurate evaluation of mineral relief is a crucial for the correct identification of minerals using the present key. For consistency, the descriptive terms "low", "medium", "high" and "very high" relief, used in this key, are defined in Table 1.2 in relation to a specified range of refractive index values.

*Table 1.2* Sub-divisions of mineral relief, used throughout the present key, in relation to defined refractive index ranges.

| Relief | | RI |
| --- | --- | --- |
| +ve | very high | > 2.000 |
| | high | 1.640–1.999 |
| | medium | 1.565–1.639 |
| | low | 1.542–1.565 |
| | Canada Balsam | 1.537–<u>1.542</u>–1.550 |
| −ve | low | 1.510–1.542 |
| | medium | 1.440–1.509 |
| | high | < 1.440 |

The "relief" of a mineral is a function of mineral refractive index, which as described in the previous section varies according to the orientation of crystallographic axes, and optic axes. The relief of a mineral is best judged in relation to the mounting medium, or surrounding crystals of known relief, such as quartz (low relief), muscovite (medium relief) or garnet (high relief).

Raising the sub-stage condenser and closing the diaphragm is recommended in order to enhance relief and make it easier to judge the relative relief of minerals. Compare for example, the bright but rather "flat" looking, poorly contrasted image of Fig. 1.7a (diaphragm fully open), with the same field of view in Fig. 1.7b, correctly adjusted. In the latter, the high relief sillimanite (centre) is easily identified, whereas in the former the relief of minerals is so subdued that it is much less easy to judge relief and thus recognise the sillimanite.

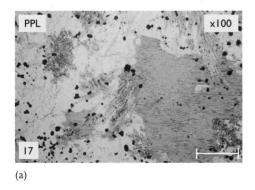
(a)

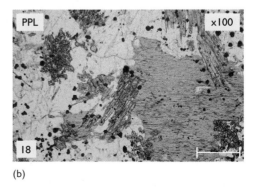
(b)

*Figure 1.7* Enhancing relief by correct adjustment of the microscope; a) poorly adjusted microscope, making it difficult to identify high relief sillimanite (centre and top); b) ideal set-up, with sub-stage condenser raised and diaphragm closed down to enhance the relief of the sillimanite. Also notice the improved definition of biotite cleavage. Bt-Sil hornfels in contact aureole of granite; Grampian Highlands, Scotland.

The "Becke line" test is a method used to determine whether a mineral is of higher or lower relief than its neighbour, or with respect to the mounting medium. Those minerals with relief higher than the mounting medium (the majority) are said to have positive relief, whilst those minerals with relief, and thus refractive indices, lower than the mounting medium are described as having negative relief. Correct evaluation of mineral relief is crucial and is the basis for many of the fundamental divisions in the various Sections of the mineral key. Although not crucial for most minerals, the determination of whether a mineral has positive or negative relief can be vital for some.

The "Becke line" method relates to the bright line that may be seen at the boundary between a mineral and the mounting medium, or between neighbouring minerals with different refractive indices. This is best seen at high magnification (x400) with the sub-stage condenser raised and the diaphragm closed. By raising the tube of the microscope, to slightly de-focus, the line is observed to move onto the material with the higher refractive index. Fig. 1.8 shows this nicely, with Fig. 1.8a showing the Becke Line at the junction between the mounting medium (left and upper) and microcline (lower right quadrant). Notice how by slightly raising the microscope tube the Becke line has moved onto the microcline (Fig. 1.8b) thus demonstrating that the microcline has higher (and positive) relief relative to the mounting medium. For further explanation of this technique see MacKenzie & Adams (1994), Kerr (1977) and Ehlers (1987a, b).

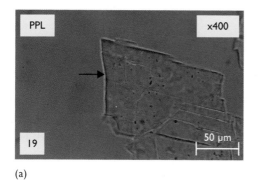 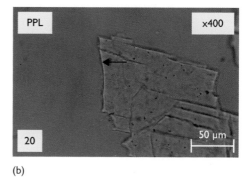

(a)                                                                 (b)

*Figure 1.8* Determination of relative relief of adjacent minerals using Becke line technique. a) Initial position with Becke Line at junction between mounting medium left and microcline (lower right quadrant); b) By raising the microscope tube to slightly de-focus, the Becke line has moved onto the higher (positive) relief microcline.

By virtue of pronounced differences in refractive indices in different crystallographic orientations, some minerals, most notably carbonate minerals (e.g. calcite and dolomite) show marked change in relief when the microscope stage is rotated. This produces an effect known as "twinkling", where the relief of individual crystals rises and falls as the stage is turned. This is often best observed by looking at two adjacent crystals of the same mineral, one showing low relief, and one high, then watching as their relative relief switches when the stage is turned. In Fig. 1.9, two images of the same field of view of an impure Cal-Qtz marble, show differences in the relief of calcite. The second image is rotated through 90° relative to the first. Notice how the calcite crystal in the centre (arrowed) changes from low/moderate relief in Fig. 1.9a to high relief in Fig. 1.9b. This reflects the full range of refractive indices from $\varepsilon = 1.486$ (medium −ve relief) to $\omega = 1.658$ (high +ve relief), but not all crystal cuts will be as favourably oriented with respect to the crystallographic axes and some may show little or no twinkling.

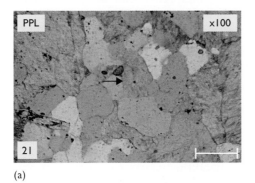 

(a)                                         (b)

*Figure 1.9* a) Impure regionally metamorphosed marble from Troms, Norway, comprised largely of calcite and quartz. The images show how individual grains of calcite show "twinkling" due to changing relief as the microscope stage is rotated; b) same field of view as in a), but rotated 90° clockwise. Note how the calcite crystal arrowed changes relief, relative to other calcite grains, and to adjacent quartz.

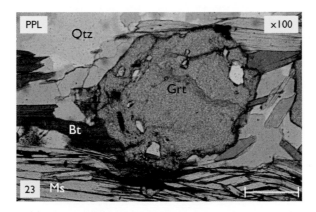

*Figure 1.10* High relief garnet porphyroblast in a matrix of moderate relief muscovite and biotite, and low relief quartz; garnet-mica schist, Ross of Mull, Scotland.

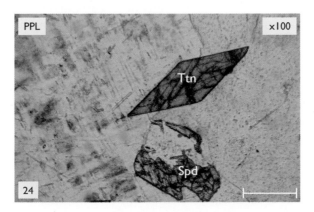

*Figure 1.11* Diamond-shaped crystal of titanite (=sphene), above, with adjacent crystal of spodumene (below) set in a background of microcline. Spodumene granite; Leinster, Ireland.

Figures 10 & 11 show two commonly encountered high relief minerals, namely garnet (Fig. 1.10) and titanite (Fig. 1.11). Fig. 1.10 is a particularly nice example, as it shows the difference between quartz (low relief), muscovite and biotite (medium relief), and garnet (high relief).

## 1.4   Mineral properties in cross-polarised light (XPL)

### 1.4.1   Birefringence and interference colours

In unpolarised light, the light oscillates in all directions perpendicular to the light (propagation) source. When the light passes through the polariser, below the microscope stage, the light becomes polarised so that beyond the polariser it only vibrates in a single plane. However, on passing through a crystal in a thin-section, the polarised light ray splits into two waves (the ordinary and extraordinary), vibrating in separate planes at right-angles to each other. These two waves propagate at different rates as a function of the crystal system, but do not interfere with each other. However, with the analyser inserted, with its polaroid filter at 90° to that of the original polariser, it produces cross-polarised light. The two out-of-phase waves are now brought back into the same plane. Since they are out of phase with each other, there is a resulting interference of the waves and this is what gives rise to the interference colours of minerals seen in cross polarised light, and a value for birefringence ($\delta$), which varies from one anisotropic mineral to the next. In essence it is a summation of the various individual wavelengths in the white light "rainbow" spectrum (ie red, orange, yellow, green, blue, indigo, violet) which are slightly out of phase with each other. In cubic minerals the waves completely cancel each other out, such that the crystals appear constantly black in cross-polarised light and are said to be isotropic.

Multiple orders (1st, 2nd 3rd etc) of interference colours are produced to give the classic Michel-Lévy interference colour chart seen in most optical mineralogy books, and also included at the back of the present mineral key. The magnitude of birefringence ($\delta$) for anisotropic minerals is directly related to the difference between maximum and minimum refractive indices, and thus optic axes. For biaxial minerals, $\delta = \gamma - \alpha$, whilst for uniaxial minerals $\delta = \varepsilon - \omega$. Although the birefringence ($\delta$) of a mineral has a defined value (or range), the interference colours observed will vary according to the thickness of the thin-section and the orientation of individual crystals. By convention, the standard thin-section thickness is 30 µm, and throughout the present mineral key the interference colours referred to are those associated with this standard thickness. If for any reason a thicker thin-section is used, higher order interference colours will be observed for a given birefringence, with the reverse true for thinner slides. For a standard 30 µm thickness thin-section, the boundaries between interference colour orders repeat at intervals approximating to $\delta = 0.018$. Hence the boundaries are at $\delta = 0.018$ (top 1st order), 0.037 (top 2nd order), 0.056 (top 3rd order), 0.074 (top 4th order) (see Demange (2012) for further detail). Since birefringence is directly related to differences in refractive indices for a given mineral, it means that minerals showing "twinkling" (e.g. calcite, dolomite) must always have high order interference colours. Note that all values for $\delta$ given in the present publication are based on data presented in Deer *et al.* (1992, 2013).

Making the correct assessment of a mineral's birefringence based on its interference colour in thin-section is crucial for the effective use of the present key. There are, however, some important points to be aware of in order to make the correct determination. Assuming that the thin-section is at the required thickness, the other factor, mentioned above, that substantially influences the interference observed for a given crystal is its orientation with respect to the thin-section cut. If the crystal is cut perpendicular, or very close to perpendicular to, the $z$-axis ($\equiv$ $c$-axis), the grain will show dark grey to black (isotropic) interference colours, by virtue of looking right down the axis. For other orientations, however, higher interference colours will be seen (unless of course it is a mineral with very low birefringence!).

The highest interference colour observed, and the one on which the birefringence of a mineral should be based, will only be seen in crystals perfectly oriented with respect to the maximum and minimum optic axes (thus giving the greatest refractive index differences). All other orientations will be oblique cuts and thus show lower interference colours. This is also important when determining extinction angle (see below).

In a thin-section the majority of crystals will be oblique cuts and thus showing a spread of colours (e.g. Fig. 1.12) slightly below the true maximum value. It is important therefore to look at the full spectrum of interference colours for the population of grains of a particular mineral, and from this identify the highest interference colour represented. If only one or two crystals of the mineral of interest are present (i.e. very small sample size), it should be realised that the birefringence could well be higher than the maximum interference colour that these few crystals might suggest. This is simply because with just a few crystals, and most likely they are oblique cuts, the peak interference colour may not be represented. In view of this, always err towards the higher side of any couplet in the key involving interference colours if you find yourself with a borderline situation. Hard minerals like corundum can sometimes be a problem as they are not always ground down to the required 30 µm thickness (by virtue of their hardness) so may give slightly elevated interference colours because of the thicker section.

Having identified the individual crystal(s) with maximum interference colour (for a particular mineral), as described in the previous paragraph, the maximum intensity of interference colour for any crystal is always seen at the mid point between extinction

*Figure 1.12* Prehnite ($\delta$ = 0.020–0.033) showing range of interference colours due to orientation of crystals in relation to thin-section cut. Highest interference colour from these crystals is 2nd Order blue (see crystal at 45° towards top left), but note all the lower 1st ord. yellow, orange, red and purple interference colours. It is very important to recognise the highest colour for a given mineral (see text for further detail). Prh-qtz vein cutting volcaniclastic sediments; Roseland, Cornwall, England.

positions. With the exception of isotropic crystals (which stay black), when the microscope stage is rotated, the crystal will be seen to pass through extinction (go black) four times in each 360° rotation (ie once every 90°). For minerals such as muscovite, with extinction parallel to the cross-wires (ie N-S and E-W), the maximum intensity of interference colour will be in the 45° position (ie NE-SW or NW-SE). See for example the muscovite and talc crystals in the combined set of images (Fig. 1.14), and the prehnite crystals in Fig. 1.12 above. Also see the radiating zeolites (Fig. 1.14, mid top row) with first order birefringence, noting that the horizontal and vertical crystals are at extinction, demonstrating they have straight extinction, whilst the crystals oriented NW-SE show peak 1st Order white interference colours.

Finally, be aware that whereas colourless minerals (e.g. muscovite) show interference colours exactly as they appear on the colour chart (Fig. 1.13a), minerals with strong colour in plane-polarised light (e.g. biotite, rutile) (Fig. 1.13b), will generally have their interference colours masked to a lesser or greater extent, making it difficult to judge the interference colours (and thus birefringence value).

To the inexperienced petrologist, it is not always easy to judge the Order that the different interference colours observed in thin-section relate to. A useful tip to help with this is to look for crystals that have a thinned or tapered edge (see Fig. 1.15). Being thinned, the interference colour represented at the edge will be less than that of the standard 30 μm thin-section. Indeed, the thinned edge often shows the full spectrum of colours going down through the orders from whatever the maximum value the crystal is showing. The multiple bands in the crystal edge zoning of the 3rd order phlogopite of Fig. 1.15c, is clearly very different to the zoning at the edge of the top 1st/low 2nd order colours of sillimanite and augite (Fig. 1.15 a, b). The sillimanite and augite examples clearly show the simple gradation down through top first order purple/red then mid first order orange/yellow/white, and just about into lower first order grey.

Anomalous Berlin blue/Prussian blue or brown interference colours are seen in some anisotropic minerals that have first order birefringence (e.g. chlorite, zoisite, melilite, vesuvianite) (see Fig. 1.16). Such colours occur in place of the greys that should be shown for a mineral of such birefringence. So why are the colours anomalous compared to those predicted by the Michel-Lévy colour chart for white light? At the start of this

(a)

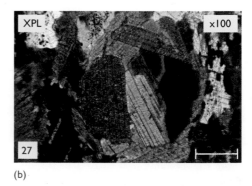

(b)

*Figure 1.13* Comparison of two minerals with similar birefringence to illustrate how the interference colours of a colourless mineral such as a) muscovite, are close to that on the Michel-Lévy colour chart, whilst those of b) biotite, in most orientations are strongly masked by the strong background colour of the mineral.

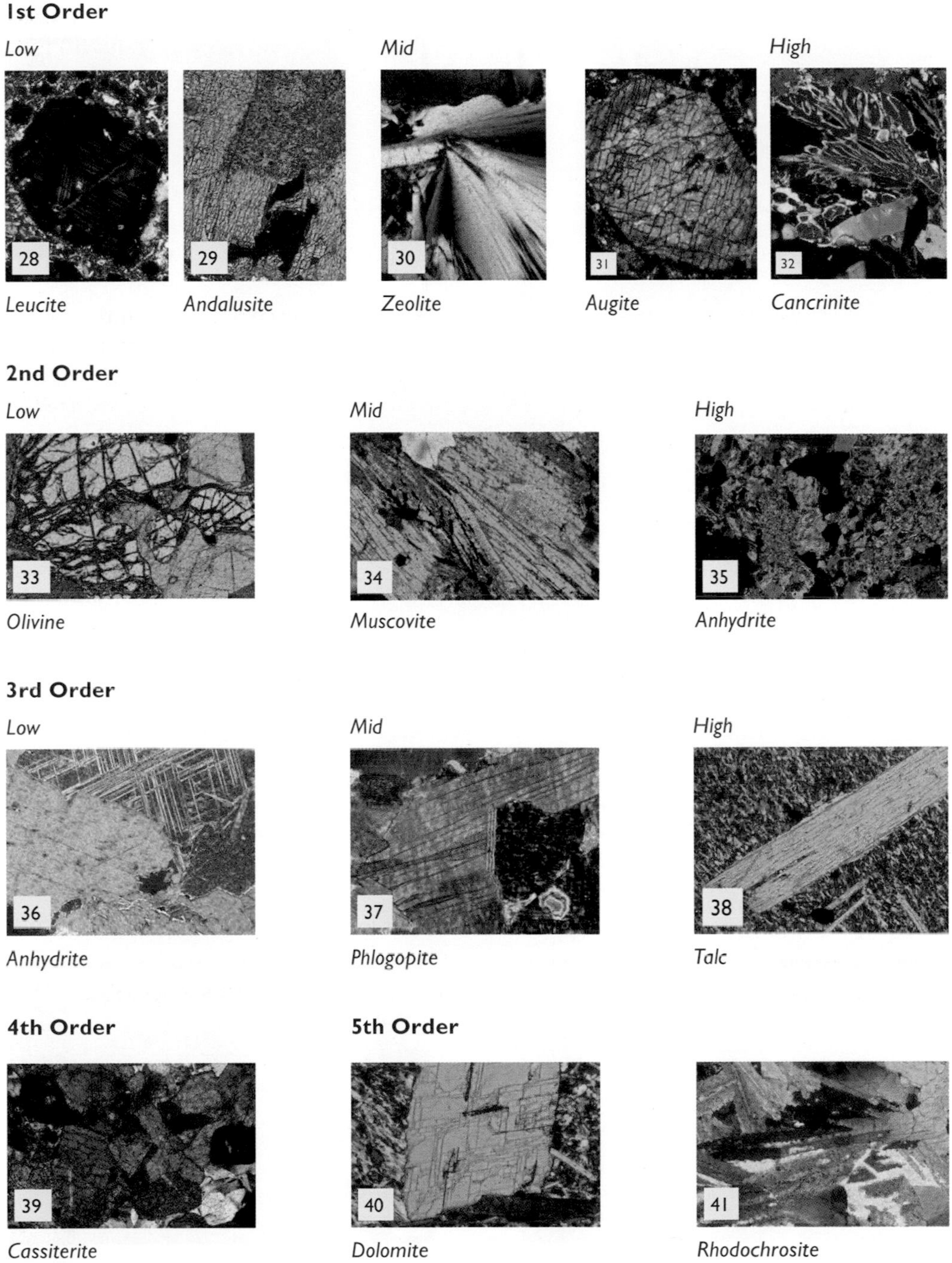
1st Order
Low
Mid
High
28
29
30
31
32
Leucite
Andalusite
Zeolite
Augite
Cancrinite
2nd Order
Low
Mid
High
33
34
35
Olivine
Muscovite
Anhydrite
3rd Order
Low
Mid
High
36
37
38
Anhydrite
Phlogopite
Talc
4th Order
5th Order
39
40
41
Cassiterite
Dolomite
Rhodochrosite

*Figure 1.14* Examples of minerals representing different interference colours according to the various levels of birefringence (see colour chart and Appendix 1 for full range of colours, and Appendix 2 for minerals listed in birefringence ($\delta$) order.

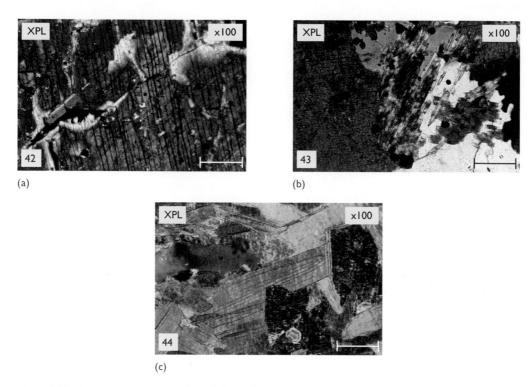

(a)

(b)

(c)

*Figure 1.15* Representative examples of thinned crystal edges as an aid to determine the maximum interference colour of a given crystal; a) augite in gabbro; Isle of Skye, Scotland; b) sillimanite in sil-kfs-hornfels; Aberdeenshire, Scotland; c) phlogopite; Bavaria, Germany. In the phlogopite example, note the multiple bands of interference colours at the thin left edge of the crystal at extinction lower centre (and other crystals). These display a sequence of interference colours covering several orders, and show the mineral is of 3rd order birefringence.

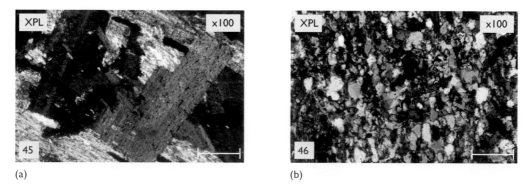

(a)

(b)

*Figure 1.16* Anomalous interference colours typical of certain minerals that based on their birefringence should show 1st Order grey-white-pale yellow interference colours, but in fact show anomalous Berlin blue/Prussian Blue or brown interference colours. a) late-stage chlorite cross-cutting schistosity in grt-mica schist; Troms, Norway; b) zoisite in Hbl-Qtz-Zo hornfels; Isle of Man, England.

section it was explained that the interference colours are produced by a composite summation of the various waves (relating to the colours of the "rainbow" spectrum, each with different wavelength) that make up white light with the analyser inserted (Fig. 1.5–21 of Kerr, 1977 gives a good illustration of this). The cause of the anomalous colours is when a mineral shows significant variation in birefringence due to different wavelengths of light. This gives rise to a lowering in intensity of certain colours, such that the combination of waves produces a colour that differs from that predicted by the Michel-Lévy chart. Although when first encountered, these unusual colours seem rather confusing, if correctly recognised, they can in fact be very useful for the identification of those few minerals that show them.

### 1.4.2  Twinning

Twinning is commonly observed in many minerals, especially those of igneous and metamorphic rocks. For some minerals the presence of twins is more or less ubiquitous, yet for other minerals twinning is rarely seen. In view of this, presence of twinning can sometimes be helpful for mineral identification.

The "simple twin" (Fig. 1.17a), is one of the commonest types of twin observed, and a wide variety of minerals show this. Such twins are particularly prevalent in phenocrysts of igneous rocks. They generally form by mirroring of the crystal structure

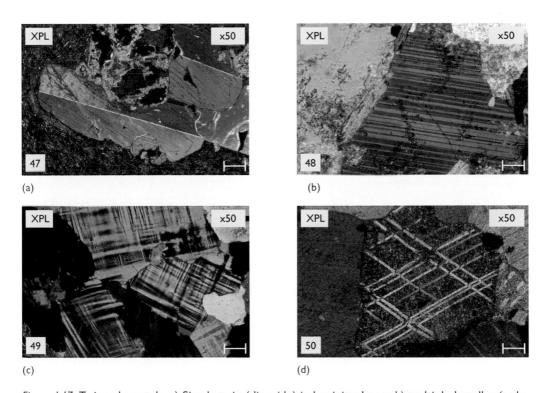

(a)                                           (b)

(c)                                           (d)

*Figure 1.17* Twinned crystals. a) Simple twin (diopside) in boninite; Japan; b) multiple lamellar (polysynthetic) twinning (plagioclase) in granite; St.Jacut, Brittany, France; c) cross-hatched twinning (microcline) in granite, Penmaenmawr, Wales; d) deformation twins (calcite) in marble; Donegal, Ireland.

each side of a particular crystallographic plane as the crystal has grown. Individual minerals have a tendency for developing twins in specific orientations relative to the crystal lattice and specific planes. Because the two halves of the twin are differently oriented within the plane of the thin-section they show different interference colours (Fig. 1.17a), and are thus a very obvious features.

Other minerals may show multiple, or polysynthetic twinning (Fig. 1.17b). This is particularly common in plagioclase of igneous rocks, yet plagioclase porphyroblasts in regional metamorphic greenschist facies metabasites do not show such twinning due to the shear stress at the time of growth. Multiple lamellar twinning forms by repeat alternations of the crystal lattice about a particular plane (010) in the crystal lattice. The regularity is such that alternate lamellae are identically oriented and thus have the same interference colour. This shows up well in the main plagioclase crystal of Fig. 1.17b, with one half of the alternating twins at extinction and the other half of the alternating twin lamellae showing showing 1st Order light grey-white interference colours.

Another distinctive type of twinning is "cross-hatched" or "tartan-twinning", which is diagnostic of microcline (Fig. 1.17c). Such twinning comprises combinations of albite-twining and pericline twinning, parallel to {010} and {100} respectively, and thus best displayed as two sets of twins at 90° to each other in planes cut approximately parallel to (001). It is outside the scope of the text to discuss feldspar twinning in any further detail, but for additional information see Deer *et al.* (1992, 2013).

A final and quite different type of twinning to consider is that of deformation twinning (Fig. 1.17d). This type of twinning is especially common in carbonate minerals (e.g. calcite, dolomite) in regional metamorphic environments and shear zones, but may also be seen in highly deformed pyroxenes and plagioclase. Unlike the primary twins described above, deformation twins are secondary in nature, having developed after original crystal growth. They form by shear along particular planes of the crystal structure and commonly give rise to more that one set of deformation twins in different orientations. Rather than intracrystalline slip, deformation twins form by twin-glide, giving rise to reorientation of the crystal lattice into the twinned position. In carbonate minerals the narrow deformation twins generally show brighter, lower order, interference colours (e.g. 2nd/3rd order) compared with the high order pastel shades (5th order) of the intervening areas of the crystal. The twins and their orientations are best observed when the main area of the crystal is placed into the extinction position.

### 1.4.3   Zoning

Various forms of zoning may be recognised in crystals, including concentric (growth) zoning, formed due to chemical changes as the crystal grew. This is particularly common in igneous phenocrysts, but can also be seen in metamorphic minerals of skarns such as vesuvianite (Fig. 1.18a) and diopside (Fig. 1.18b). Other types of zoning include crystallographically controlled zoning, such as the distinctive "hour-glass" zoning of chloritoid (Fig. 1.18c), and "sector-zoned" titanaugite (Fig. 1.18d) and cordierite (Fig. 1.18e). A final type is the well known "chiastolite-cross" (Fig. 1.18f) of andalusite (var. chiastolite) often seen in metapelitic rocks within the contact aureole of granites. Although most types of zoning are not unique to specific minerals, they are sufficiently distinctive for a few minerals that they can be a useful aid to identification.

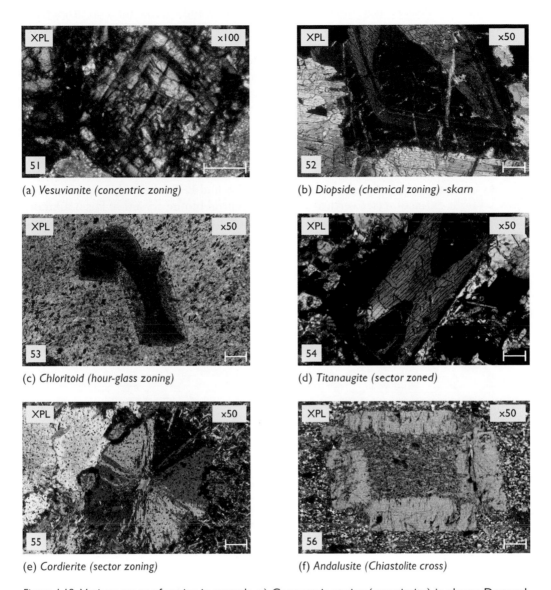

(a) *Vesuvianite (concentric zoning)*

(b) *Diopside (chemical zoning) -skarn*

(c) *Chloritoid (hour-glass zoning)*

(d) *Titanaugite (sector zoned)*

(e) *Cordierite (sector zoning)*

(f) *Andalusite (Chiastolite cross)*

*Figure 1.18* Various types of zoning in crystals. a) Concentric zoning (vesuvianite) in skarn; Donegal, Ireland; b) chemically zoned diopside in calc-silicate skarn; Loch Ailsh, Assynt, Scotland; c) hour-glass zoning (chloritoid [*ottrelite*]) in chloritoid phyllite; Ottre, Belgium; d) sector zoned titanaugite in nepheline syenite; Saxony, Germany; e) Sector zoning/sector trilling (cordierite) in cordierite-andalusite slate, Connemara, Ireland; f) chiastolite cross (andalusite) in cordierite-andalusite slate; Connemara, Ireland.

### 1.4.4   Extinction angles

When describing birefringence, in 1.4.1 above, it was stated that when the microscope stage is rotated in cross polarised light, the crystal under consideration will go into extinction four times in a complete 360° rotation (ie once every 90°). Whilst this is true of all anisotropic minerals, the position at which extinction occurs varies from one mineral to the next as a function of crystal system and specific mineral properties. The extinction angle also varies from one crystal to the next as a function of crystal orientation in the plane of the thin section.

It is easiest to define extinction in relation to elongate crystals and those with well developed cleavage and/or crystal faces (seen as flat edges in thin-section). If the crystal goes into extinction with the flat edge of crystal length in a position parallel with the N-S (vertical) or E-W horizontal position it is said to have straight or parallel extinction (i.e. 0°) (see Fig. 1.19). If the microscope eye-piece (ocular) contains N-S and E-W oriented cross-hairs this can be accurately determined; if not judgement by eye will be required. Similarly, length-section crystals showing a good cleavage trace (usually parallel to crystal edge), can also have their extinction angle easily determined. If the crystal goes into extinction with the cleavage trace N-S and E-W oriented, it too is said to show straight extinction (see vertically oriented central biotite crystal of Fig. 1.13a, and stilpnomelane of Fig. 1.19a). For small elongate grains with no obvious cleavage and poorly defined crystal edges, the best way to assess the extinction angle is to line up the

(a)

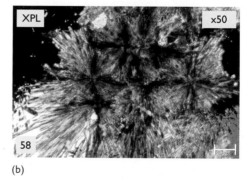

(b)

*Figure 1.19* Minerals showing straight extinction. a) Stilpnomelane (note lower left E-W crystal and central N-S crystals at extinction), in metamorphosed siliceous ironstone; Laytonville Quarry, California, USA; b) Radiating tourmaline crystals in Luxullyanite; Cornwall, England. Note: Those crystals in the E-W and N-S positions are at extinction, thus giving a series of black crosses; one cross for each radiating cluster.

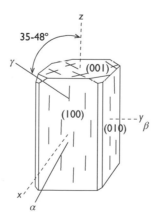

*Figure 1.20* Augite crystal showing extinction angle based on the angular difference between $\gamma$ optic axis and z- crystallographic axis.

length of the crystal N-S. If it shows extinction when aligned this way, it can reasonably be interpreted as being straight extinction.

For minerals with straight (=parallel) extinction, there is parallelism between the crystallographic axes and optic axes (vibration directions). However, minerals showing inclined extinction (e.g. clinopyroxenes, clinoamphiboles), have varying levels of angular difference between the optic axes and the crystallographic axes. Augite for example (Fig. 1.20), shows a maximum difference between the $\gamma$-axis (optical axis) and the $z$-axis (crystallographic axis), $\gamma$ :$z$ of 35–48°, in side-sections cut parallel to the $x$-$z$ plane.

If the crystal being observed in thin-section is not roughly parallel with the plane containing $x$–$z$ axes, (010), the angular difference $\gamma$:$z$ will be less than the maximum extinction angle (i.e. it gives an apparent $\gamma$:$z$ angle). It is therefore essential when determining the extinction angle for a mineral that the value is measured for a representative sample of crystals to find out what the maximum (and thus true) extinction angle is. Those crystals with the favourable orientation, containing the $z$-axis, will be those showing the highest interference colours, and conversely those showing the lower or lowest interference colours should be avoided. The significance of this was also discussed in relation to interference colours and determination of birefringence (see 1.4.1, above). The present mineral key uses the broad distinction between straight and inclined extinction minerals as one of the fundamental divisions (Section 2). Within Section 2 many of the couplets then rely on the precise value for extinction angle, hence the importance of selecting the right crystals for this determination. Accurate measurements can be made by positioning the crystal in the centre of the field of view, and lined up with the cleavage trace or crystal edge N-S.

Once done, the orientation of the microscope stage is recorded using the fixed vernier scale present on one edge. Having noted a figure for the first position, the stage (and crystal) are then rotated left and right from the N-S position to the find the point of extinction. The new position of the microscope stage is then recorded, and the difference between the two gives the extinction angle. By convention, this is always quoted as the lower angle. If for example, a mineral went extinct when angled 31° to the left of the N-S position, it will also show extinction at 59° to the right, since extinction happens every 90°. In this case, the extinction angle is correctly recorded as 31°. If there are enough suitable crystals in the thin-section, the present author recommends checking the extinction angle on at least five suitably oriented crystals, and preferably

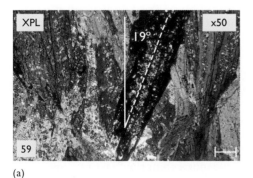

(a)                                                        (b)

*Figure 1.21* Minerals showing inclined extinction. a) actinolite in albite-epidote hornfels; locality unknown, and, b) plagioclase in troctolite; Sierra Leone.

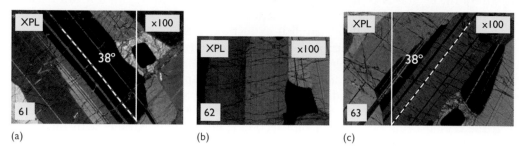

*Figure 1.22* central image (b) is plagioclase with optimum orientation as shown by equal illumination in N-S position; a) and c) are the same plagioclase rotated left to the position with one set of twins at extinction, and right to the position where the other set goes into extinction. Extinction angles have been measured and displayed. The average of the two gives the extinction angle for the crystal. Note that such averages should only be calculated if the left and right angular values for extinction are broadly similar (which indicates the crystal was ideally oriented). Anorthitic plagioclase in anorthosite; Rhum, Scotland.

ten. Fig. 1.21 shows examples of inclined extinction for actinolite (Fig. 1.21a) and plagioclase (Fig. 1.21b).

When measuring extinction of multiple lamellar twinned plagioclase, look for crystals with equal illumination when in N-S position relative to cross-wires (Fig. 1.22b), then rotate left and right in turn (Fig. 1.22a-c) to measure extinction angle of each twin set. If the crystal is to be regarded as reliable for extinction angle determination, the value obtained turning left should be the same or very close to the value obtained when turning right. In this example the extinction angle is 38° indicating the plagioclase in question is a Ca-rich labradorite ($An_{68}$), bordering on bytownite. When the extinction position bisects the angle between two cleavages, as typically occurs in end-sections of pyroxenes, it is said to be symmetrical extinction.

### 1.4.5   Length-fast and length-slow crystals

In 1.4.1, it was explained how white light is polarised into a single plane when it passes through the lower polariser. This polarised light hits the underside of the thin-section as a wave vibrating in a single plane. However, immediately on entering the crystal the light splits into two rays (ordinary ray and extraordinary ray), vibrating in separate planes at right-angles to each other. These rays move at different velocities as a function of the crystallographic structure and atomic arrangement. They are generally referred to as the "fast ray" and "slow ray". When the analyser is inserted (polariser at 90° to that of lower polariser), because one ray is moving faster than the other, they are out of phase with one another, and combine to produce the interference colours of cross-polarised light. This is true for anisotropic minerals, but for cubic minerals, which have identical atomic arrangement in all directions, the rays move at the same rate and thus combine perfectly to cancel out and give darkness (ie the mineral is isotropic).

The two rays described above for anisotropic minerals have different refractive indices, the "slow ray" always having a higher refractive index than the "fast ray". Because of these differences, determination of the fast and slow vibration directions has proved to be a very useful way of identifying certain minerals that are otherwise quite similar to each other. The method involved with this simple but effective technique will now be described.

Unless adjusted for some reason, remember that the polarisers of the microscope should be aligned with one polariser parallel to E-W and the other parallel to N-S. By inserting the analyser, so that the thin-section is viewed in cross-polarised light, the stage is rotated to find the extinction position of the mineral (see 1.4.4). Bearing in mind the orientation of the polarisers, the positions of extinction (at 90° intervals), give the orientations of the vibration planes of the fast and slow rays. When the mineral goes black, the vibration directions of the rays are essentially parallel to the polarisers. At this stage, however, there is no way of knowing which orientation represents the fast ray, and which the slow ray. For minerals with straight extinction, the rays are parallel with the crystallographic axes, whereas in the case of inclined extinction, the rays (optic axes) are oblique to the crystallographic axes (see 1.4.1 above).

Having established the orientation of the rays based on mineral extinction positions, the mineral is then rotated to a position at 45° to the extinction position (ie bisects the 90° angle between extinctions). At this position the crystal will show maximum intensity of interference colour. If we take the case of a mineral showing straight extinction, the crystal would be rotated to either the NW-SE or NE-SW position.

The next step is to insert a mica ($\lambda/4$) plate (retardation $\Delta = 150$ nm), a gypsum ($\lambda$) plate ($\Delta = 550$ nm) or a quartz wedge (typically ranges from $\Delta = 0$–50 nm up to $\Delta = 2200$ nm). The retardation due to the quartz wedge increases progressively with increasing thickness of the wedge (see Kerr, 1977, for further detail). The choice of accessory plate to use depends on the birefringence of the mineral under investigation. The mica plate is typically used for those minerals with very low anisotropy (low 1st order interference), the gypsum ($\lambda$) plate for minerals with greater anisotropy (e.g mid first order to second order). The quartz wedge, because of its tapered nature of gently increasing thickness produces interference colours from low first order to fourth order. It is a good general purpose accessory plate, but is especially useful for determining the orientation of fast and slow rays in minerals with high order interference (e.g.carbonate minerals).

It is standard for most microscopes to have the accessory plate insertion position as from NW or SE. Fig. 1.23 gives a schematic representation of a typical gypsum ($\lambda$) plate, with "fast" and "slow" orientation indicated (Fig. 1.23a). On insertion of the plate from the NW, a crystal oriented length-wise NW-SE, would be interpreted as "length-fast" if the interference colour increased on insertion of the plate (Fig. 1.23b). Conversely, a crystal is "length-slow" if it shows a decrease in interference colour when the plate is inserted with the crystal length aligned NW-SE (Fig. 1.23c). In each case, if the crystal is turned through 90°, the colour reverses (Fig. 1.23b, c).

It is straightforward to explain why the particular colours occur, as they are simply the combination (addition or subtraction) of the retardation value ($\Delta$) of the mineral, based on its birefringence (Appendix 1), with that of the accessory plate used. If we consider the case of an andalusite crystal showing 1st order white interference colours ($\Delta = c.200$ nm) in XPL, when the gypsum plate ($\Delta = 550$ nm) is inserted (see Fig. 1.24b), the combined $\Delta$ value obtained will be either an addition ($\Delta = 550 + 200 = 750$ nm;

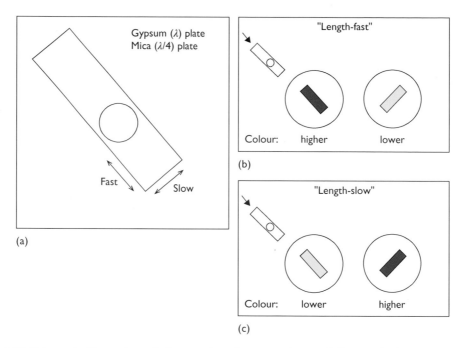

*Figure 1.23* Schematic illustration of changing interference colours of "Length-fast" and "Length-slow" minerals when a gypsum ($\lambda$) plate ($\Delta = 550$ nm) is inserted. a) orientation of accessory plate at point of insertion varies between microscopes, many have the plate inserted NW-SE. Fast and slow vibration directions indicated; b) for length-fast crystals, when the plate is inserted as shown, with the crystal aligned with its "fast" orientation parallel to the "fast" direction of the plate (left), the interference colours of the mineral will be observed to increase. With the crystal oriented with its "fast" orientation (length) at 90° to the inserted plate (right), the interference colours will be lower; c) length-slow crystals show the reverse relationship, with crystals aligned with their "slow" direction parallel to the "fast" direction of the inserted plate showing a decrease in interference colours, whilst at 90° to this the crystal shows higher colours. Note that the blue and yellow used in the diagram are representative of the change for crystals with 1st order grey/white interference colours when the gypsum plate is inserted, they do not necessarily reflect the various colours that may be seen at other values of birefringence and when using other accessory plates.

2nd order blue-green) when the fast ray of the plate and that of the crystal are aligned, or subtraction when the slow ray of the crystal aligns with the fast ray of the plate (ie $\Delta = 550 - 200 = 350$ nm; 1st order yellow).

Fig. 1.24 illustrates some representative examples of length-fast and length slow crystals. Notice how the two intergrown chloritoid crystals of Fig. 1.24a have the NW-SE oriented crystal showing the high colour and the NE-SW crystal showing the low colour. Compare this with the schematic representation in Fig. 1.23b for length-fast crystals. Also note the three crystals of andalusite (var. chiastolite) in Fig. 1.24b, the NW-SE oriented crystal (centre) showing the high colour, the NE-SW crystal (right) showing the low colour, and the horizontal crystal (left edge), oblique to insertion direction (and at extinction when plate was inserted) showing a colour intermediate between the two, close to that of the gypsum plate. Tourmaline (Fig. 1.24c) is another commonly encountered length-fast crystal, in this case producing a combined 3rd order green colour.

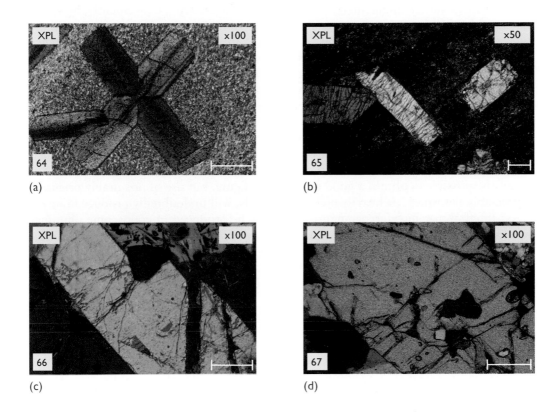

*Figure 1.24*   Representative examples of length-fast and length-slow crystals; λ-plate inserted from NW in all cases. a) Chloritoid (length-fast), in chloritoid phyllite; Tarkwa, Ghana; b) Andalusite (length-fast) in andalusite slate; Skiddaw Granite aureole, Cumbria, England; c) Tourmaline (length-fast), in tourmaline-quartz vein rock; Cornwall, England; d) Staurolite (length-slow), in staurolite schist; Snake Creek, Mt.Isa region, Queensland, Australia.

The staurolite (Fig. 1.24d), showed 1st order grey/white interference colours prior to insertion of the gypsum plate. With the plate inserted, and with length oriented NW-SE, it now shows its length-slow character, with lower colours (1st order yellow-orange) compared to the gypsum red, exactly the opposite of chloritoid and andalusite above.

### 1.4.6   Interference figures

All previously described thin-section properties of minerals are observed in what is termed orthoscopic illumination. This involves the various light rays moving in parallel paths upwards through the lens system of the microscope (and through the thin-section) up to the eye-piece where an image of the thin-section field of view is projected. By contrast, interference figures are produced using conoscopic illumination. This involves light being converged to a point within the thin-section, and beyond this, the light leaves the thin-section as a slightly conical beam. A close working distance between objective lens and thin-section is required for the divergent conical beam of light to form an interference figure on the top curved surface of the objective lens. This can be viewed by taking the eye-piece out and looking straight down the microscope tube at the tiny image on the upper surface of the objective lens. More conveniently, however, most good polarising microscopes will have a Bertrand lens fitted at some point up the microscope tube. Inserting (or flipping) this lens in position projects the interference image from the top of the objective lens up to the eye-piece for the viewer to observe. The interference figure comprises a series of dark lines ("isogyres"), relating to the optic axes; a simple cross for uniaxial minerals and two curved black lines for biaxial minerals. For further detail on the topic of interference figures, the interested reader is referred to Kerr (1977), Ehlers (1987a,b) and Demange (2012).

To obtain a good interference figure, it is important to select an end-section (or side-section) crystal with low interference colour (1st order very dark-grey to black), as such crystals are cut close to parallel with crystallographic axes. Even with this, it may not always be possible to obtain a good interference figure, but use of unsuitably oriented, appreciably deformed, or heavily included crystals, will undoubtedly produce failure.

As stated above, interference figures can only be produced in conoscopic illumination (= convergent polarised light). This is achieved by raising the sub-stage condenser to just below the microscope stage, then flipping the front lens of the sub-stage condenser "in". This will give convergent light in a small central area of the field of view of the crystal of interest. This is also useful when working at high magnification and needing increased illumination.

Working in cross-polarised light (essential for interference figures), it is best to start at low magnification (i.e ×10 eye-piece and ×10 objective = ×100 magnification), then work up to progressively higher magnification, re-focusing each time. Typically through the ×20 objective (×200 magnification) and finally ×40 objective (×400 magnification). Once at ×400 magnification, insert the Bertrand lens, and if the crystal is close to the ideal orientation and not heavily included or deformed, it should be possible to obtain an interference figure.

For best results, as well as being favourably oriented to the optic axis, the single crystal of interest needs to fill the entire field of view. This is why it is usually necessary to work at ×400 magnification. Another reason for working at high magnification is because of the requirement to have a very short working distance between the objective lens and the thin-section. This is needed so that on leaving the thin-section, the diverging cone of light is still reasonably tight, and so forms a sharp interference figure on the upper surface of the objective lens. The Numerical Aperture (NA) of the objective lens (usually marked on the side of the lens, along with magnification), is important, as it directly relates to the maximum angular spread of light rays that can be transmitted. Generally speaking, the higher the magnification of the objective, the higher the NA value of the lens, but as some high magnification lenses are designed for greater working distance they will have lower NA. It is advisable to check the lens you are using. The images shown in Fig. 1.25 and Fig. 1.26, below, were all taken using a ×40 objective, with NA = 0.65. This is a fairly typical NA value for a lens of such magnification. If the crystal of interest is reasonably large and inclusion free it may be possible to obtain a reasonable optic figure at ×200 magnification (×20 objective), although from the author's experience best results are usually obtained at higher magnification (×400).

### Uniaxial figures

If in the plane of the thin-section, the crystal has been cut exactly perpendicular to $z$-axis ($\equiv c$-axis) such as the example of quartz in Fig. 1.25a, a perfect interference figure will be formed. For uniaxial minerals, this is a "cross", with dark lines (isogyres)

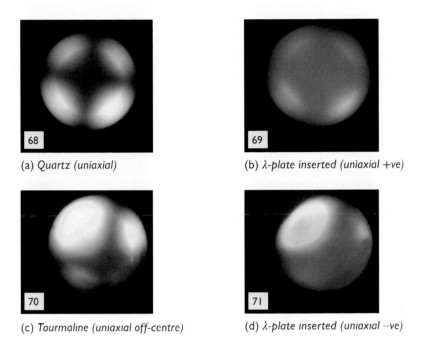

(a) *Quartz (uniaxial)*

(b) *λ-plate inserted (uniaxial +ve)*

(c) *Tourmaline (uniaxial off-centre)*

(d) *λ-plate inserted (uniaxial −ve)*

*Figure 1.25* Uniaxial interference figures (magnification x400 in all examples); a) The classic uniaxial cross in a single large quartz crystal cut perpendicular to the z-axis (≡ c-axis) (see earlier section), with Bertrand lens inserted; b) is the same view as (a), with Bertrand lens in, but now with the addition of a sensitive tint (gypsum) λ-plate inserted from the NW orientation [quartz wedge could also be used]. The lower colours in the NW and SE quadrants and band of higher colours NE-SW demonstrates that the quartz is uniaxial positive; c) Tourmaline (end-section) view with Bertrand lens inserted, showing uniaxial cross (slightly off-centre); d) is the same view, with additional insertion of gypsum λ-plate from NW. The colour arrangement is the reverse of that seen for quartz in (b), and shows that tourmaline is uniaxial negative.

oriented N-S and E-W, and crossing in the centre to divide the field of view into four quadrants. More usually however, the crystallographic axes are not so perfectly aligned and the interference figure will be off-centre, as in the tourmaline of Fig. 1.25c. Even this example, however, would be regarded as good. Most commonly the alignment is much less good, and what should appear as a cross in the case of uniaxial minerals, has the centre of the "cross" at the edge or outside the field of view. In such cases, when the microscope stage is rotated, each arm of the "cross" rotates into the field of view and out again in sequence (see Fig. 1.6–10 of Ehlers, 1987a). Such situations are normal and still of use, since by rotation of the stage the uniaxial character can still be confirmed.

Having confirmed the uniaxial nature of the mineral, the next step is to determine the optic sign. This is done by inserting the gypsum "red" plate or quartz wedge. The choice depends on the appearance of the interference figure, which will vary according to the birefringence of the mineral. If the interference figure is essentially a black cross, with white quadrants, as typifies a 1st Order birefringence mineral such as quartz (Fig. 1.25a), the gypsum (λ) plate is inserted. With insertion from NW-SE, if the NW-SE quadrants go to lower colours (1st order yellow-orange),whilst the NE-SW quadrants change to higher colours (2nd order blue or blue-green), the mineral is uniaxial positive (see Fig. 1.25b). If the reverse is true, the mineral is uniaxial negative.

As previously discussed in 1.4.5, above, the colour formed is essentially the gypsum ($\lambda$"red") plate $\Delta$ value (550nm), plus or minus the $\Delta$ value relating to the birefringence of the mineral. In off-centre examples, by knowing which quadrant of the cross is in the field of view, based on which way the arms of the cross rotate in, the optic sign (+ve or −ve) can still be determined according to whether the quadrant turns blue or else orange-red (see examples in Fig. 1.25).

### Biaxial interference figures

The approach to selecting an appropriately oriented crystal is the same as that described for uniaxial minerals. In general end-sections close to perpendicular with the $z$-axis of the crystal and showing low birefringence are the best choice, and once again, deformed or heavily included crystals are best avoided.

Having selected the crystal of interest, the same procedure as that described above, for uniaxial crystals, is undertaken. Assuming the crystal has a favourable orientation, when the Bertrand lens is inserted, if a well-centred interference figure is produced, the pattern observed will either be of two boomerang-like curved isogyres arching towards the centre of the field of view (Fig. 1.26c), or else something that looks very like a uniaxial cross (Fig. 1.26a). The latter occurs in cases where the 2V angle (optic axial angle) is very low (0–10°), whereas the latter forms with higher 2V values.

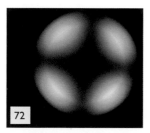

(a) Phlogopite (biaxial) $\beta = 1.597$

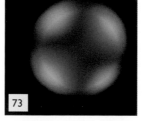

(b) Phlogopite with $\lambda$-plate inserted (biaxial −ve) (2V=0–10°).

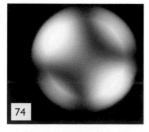

(c) Muscovite (biaxial) $\beta = 1.596$

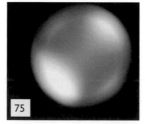

(d) Muscovite with $\lambda$-plate inserted (biaxial −ve)

Figure 1.26 Biaxial minerals (magnification x400 in all examples); a) Phlogopite with low 2V angle (0–10°) shows what at first glance appears to be a "uniaxial cross", but on rotation the two isogyres fluctuate in and out slightly revealing that the crystal is actually a biaxial mineral with very low 2V; b) is the same field as shown in a), but with the $\lambda$-plate inserted. Higher colours NW-SE and lower colours NE-SW demonstrate that phlogopite is biaxial −ve; c) Muscovite with two nicely defined isogyres arching half-way across the field of view with a boomerang-like angular curvature indicative of 2V c.30–45°; d) the same field of view as c), with $\lambda$-plate inserted. Higher colours NW-SE and lower colours NE-SW demonstrate that phlogopite is biaxial −ve.

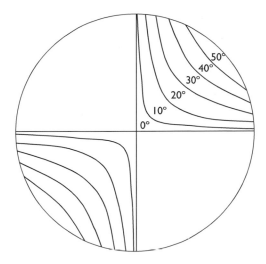

*Figure 1.27* Schematic representation of biaxial mineral 2V angles for comparison to aid estimation of 2V for an unknown mineral. $\beta = 1.60$, N.A. $= 0.85$ (diagram created based on information in Ehlers, 1987a).

*Table 1.3* Maximum 2V value (at edge) where both isogyres can be observed in the field of view, as a function of Objective lens Numerical Aperture (NA) and average refractive index of the mineral ($\beta$).

|             | $\beta = 1.4$ | $\beta = 1.5$ | $\beta = 1.6$ | $\beta = 1.7$ | $\beta = 1.8$ | $\beta = 1.9$ |
|-------------|---------------|---------------|---------------|---------------|---------------|---------------|
| NA $= 0.85$ | 74°           | 68°           | 64°           | 60°           | 56°           | 53°           |
| NA $= 0.65$ | 55°           | 51°           | 48°           | 45°           | 42°           | 40°           |

In cases that look like a cross, to prove that the crystal under consideration is a biaxial mineral with low 2V rather than a uniaxial mineral, rotation of the microscope stage should provide the necessary proof. If the stage is rotated, slight movements in and out will prove the existence of two biaxial isogyres, whereas if there is no such movement the crystal under examination is uniaxial.

Assuming that an appropriate crystal is available, quantifying the 2V angle for minerals that have 2V < c.60° is relatively straightforward, as there is a regular relationship between the position and curvature of the isogyres (Fig. 1.27). This is based on the extent to which the apex of the curve arches towards the centre of the field of view, and how close the isogyres are to the edge. However, for minerals with 2V >60°, the angle is much harder to determine because in most cases, both isogyres will not usually be in the field of view together.

Knowing the Numerical Aperture (NA) value of the objective lens being used is especially important when determining the 2V value for biaxial minerals (see below), as it influences the maximum 2V value where both isogyres are present in the field of view. In fact, it is slightly more complicated than this, because the $\beta$-index (intermediate refractive index of the mineral) also has an effect. In simple terms, the higher the NA value of the lens, and the lower the $\beta$ value of the mineral, the greater the maximum value of 2V with both isogyres observable in the field of view. Table 1.3 (above) provides a simplified summary of this for objective lenses with NA = 0.85 and 0.65, and a selection of $\beta$ values (information derived from Fig. 1.8–18 of Ehlers, 1987a).

For a helpful tabulation of biaxial minerals, arranged in ascending order of 2V angle, and also listing corresponding $\beta$ values for each mineral, Charts 5 & 6 in Ehlers (1987a) are extremely useful. Likewise, in Kerr (1977), Tables 10–9 and 10–10 and Charts A–G provide similar data.

There are advanced techniques described by Ehlers (1987a) for estimation of the 2V angle for minerals with high 2V (isogyres outside the field of view), but these go well beyond what can reasonably be addressed in the present publication. In the present mineral key, whenever 2V angle forms the final discriminator for two closely similar minerals it is often just a case of deciding whether the 2V angle is low or high that is needed, so obtaining a more precise estimate is not usually crucial.

# Chapter 2

# Key to rock-forming minerals in thin-section

## 2.1 TO DECIDE WHICH SECTION

To use the key, firstly define whether the mineral is colourless or coloured in plane-polarised light, and whether it shows cleavage. If it is a cleaved mineral, it is necessary to define the number of cleavage traces present. Finally, it is necessary to determine whether the mineral has straight or inclined extinction. Once defined, use the list below to decide which Section to use to key-out the mineral.

Section 1:   2 (or 3) cleavage traces observed pg.35

Section 2:   1 cleavage trace,
             INCLINED extinction pg.53

Section 3:   1 cleavage trace, STRAIGHT
             extinction, COLOURLESS pg.71

Section 4:   1 cleavage trace, STRAIGHT
             extinction, COLOURED pg.87

Section 5:   Imperfect cleavage,
             parting or arranged fractures pg.97

Section 6:   0 cleavage, COLOURLESS pg.111

**Section 7:    0 cleavage, COLOURED** 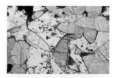 pg.131

**Section 8:    OPAQUE minerals** 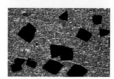 pg.145

## 2.2   Symbols & abbreviations

(□) = end-section
( ▐ ) = side-section
(...) = small crystals (eg fine granules, inclusions); cleavage or other diagnostic
    properties not evident.
// = parallel
α, β, γ = optic axes (and refractive indices)
δ = birefringence value
Δ = retardation value
+ve = positive
- ve = negative
2V = optic axial angle
birefr. = birefringence
brn. = brown
c'less. = colourless
clv. = cleavage(s)
grn. = green
interf. = interference
L/W = length/width ratio of prismatic crystals
met. = metamorphism
Mod = moderate
NA = Numerical Aperture (of objective lens)
ord. = order
pleochr. = pleochroism
RI = refractive index
RL = reflected light
rlf. = relief
$x$, $y$, $z$ = crystallographic axes
yell. = yellow

This key covers more than **150 of the most commonly encountered rock-forming minerals,** plus a few rarer but noteworthy minerals (e.g. indicative of P-T conditions, ore deposits or fluids). In some cases it may not be possible to key out specific minerals, since several minerals have very similar properties. In such cases the use of a micro-probe or other analytical techniques may be necessary to determine the mineral.

# 2 (or 3) cleavage traces

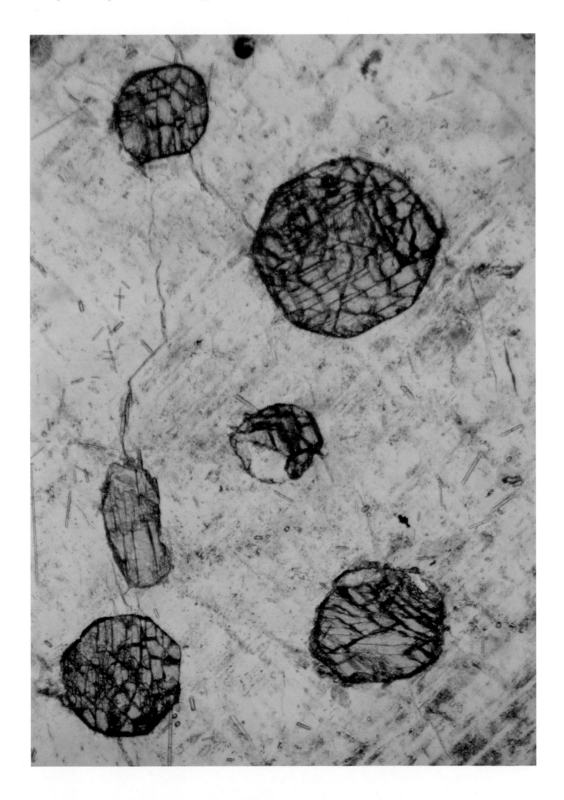

| 1 | 3 CLEAVAGE TRACES | 2 |
|   | 2 Cleavage traces | 4 |

| 2 | Isotropic | 3 |

Non-isotropic (c'less; 2 clv. at 93°; strong parting // 100) **(1a)**   **SPODUMENE (□)**

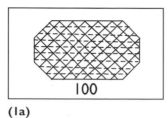

**(1a)**

Spodumene in granite; Leinster, Ireland

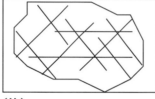

3   Colourless (or pale purple); (3 cleavage traces at 60–70°) **(1b)**   **FLUORITE**

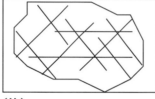

**(1b)**

Fluorite (arrowed) in hydrothermally altered garnet-rich quartzo-feldspathic metavolcanic rock; Nisserdal region, Telemark, Norway.

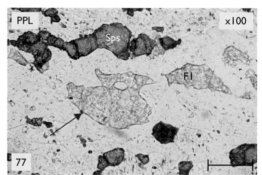

Yell.-brn. or brown (1 strong clv. + intersecting partings) **(1c)**   **SPHALERITE**

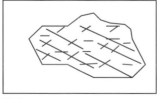

**(1c)**

Sphalerite in Zn-ore deposit; Zinkgruvan, Sweden.

4    2 CLEAVAGE TRACES at 60° (≡ 120°)    5
     2 CLEAVAGE TRACES at > 60°(< 120°)    14

5    COLOURLESS (or very pale green or brown)    6
     COLOURED (good pleochroism)    9

6    Top 1st ord./2nd ord. birefr.; colourless to pale green.    7
     3rd order birefr., pale yellowy brown.    **GRUNERITE(□)**

7    Symmetrical extinction "diamond"-shaped (**1d**)    8
     Symmetrical extinction "truncated diamond"-shaped (**1e**)
     **CUMMINGTONITE(□)**

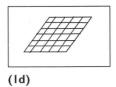

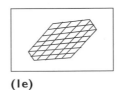

**(1d)**                    **(1e)**

8    Biaxial +ve (meta-ultramafics, other Mg-Fe rich rocks)    **ANTHOPHYLLITE(□)**

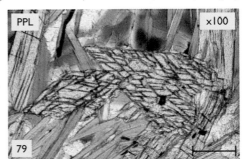

Anthophyllite (end-section) in anthophyllite-
bt-grt gneiss; Finland.

     Biaxial –ve (calc-silicate rocks, other Ca-Si-Fe rich rocks)    **TREMOLITE(□)**

Tremolite (end-section) in calcareous semi-
pelitic schist; Mangkuma-Ketempe, Ghana.

9   Blue/violet pleochroism, 1$^{st}$ ord. yell./orange birefr.                    10
    Not blue/violet pleochroism                                                   11

10   Colourless/lavender blue pleochroism;          **GLAUCOPHANE/CROSSITE(□)**

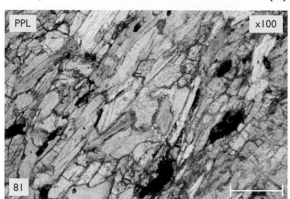

Glaucophane (end and side-sections) in glaucophane schist (blueschist); Ile de Groix, Brittany, France.

Intense prussian blue-indigo blue-green pleochroism.          **RIEBECKITE(□)**

Riebeckite (end-section) in sodic granite; Drammen, Norway.

11    Green, yellow-green, blue-green pleochroism                                    12
      Brown, greenish brown, yellow-brn, red-brown pleochroism          13

12    Typical amphibole 6-sided diamond-shaped end-section (**1e**) well developed cleavage. Green, blue-grn. to yellow green pleochroism (mafic rocks).

**HORNBLENDE / ACTINOLITE(□)**

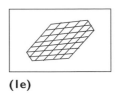

**(1e)**

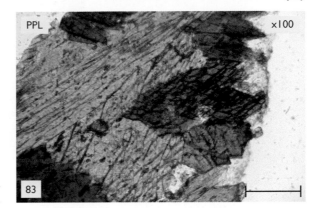

Actinolite (end-section) in act-ab-czo hornfels; locality unknown.

Equal-sided hexagon (**1f**), weak cleavage; pale green to slaty blue pleochroism (Al-rich pelite).

**CHLORITOID(□)**

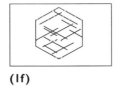

**(1f)**

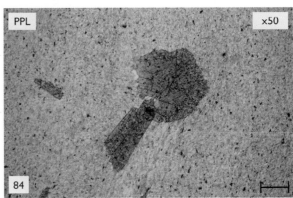

Chloritoid (end-section) in chloritoid phyllite; Tarkwa, Ghana.

13    Pale yellow/yell.-brn. to dark chestnut brown pleochroism. 2V = 44–90°. A common mineral found in a various volcanic rocks from basalts and andesites to trachytes.    **brown HORNBLENDE (Oxo-HORNBLENDE)(□)**

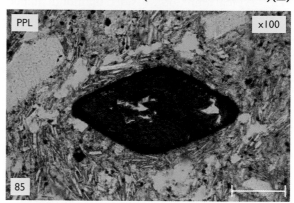

Oxo-hornblende phenocryst in andesite; Plomb du Cantal, France.

*Note: The black iron-oxides often seen around the edge of Oxo-hornblendes result from resorption effects. Such crystals probably crystallised as typical magnesio or ferrohornblende but were transformed during later stage oxidation (Deer et al., 2013).*

Brownish yellow/yell.-brn to red-brn. pleochrosim. 2V = 74–82°. Alkaline volcanic rocks only.    **KAERSUTITE(□) (previously called BARKEVIKITE)**

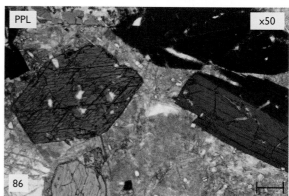

Kaersutite phenocrysts in syeno-gabbro ("lugarite"); Lugar Sill, Lugar, Ayrshire, Scotland.

*Note: Most hornblendes are pleochroic in shades of yell.grn.-green-blue-grn. However, high temperature hornblendes may be green to brown, or brown pleochroic. The amphibole KAERSUTITE, found in alkaline volcanic rocks, with yell.brn.-red-brn.-greenish brown pleochroism is difficult to distinguish from brown hornblendes. Also consider AENIGMATITE, a very dark brown, high relief almost isotropic mineral of nepheline and sodalite syenites. Its extinction angle is 66°.*

14   **2 CLEAVAGES** at c.67–75° (≡ 113–105°)                          **15**
     Cleavages at > 75°                                               **19**

15   Isotropic (usually 2 feint clv. but sometimes 3 at 60–70°) **(1g)**    **FLUORITE**

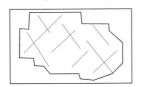

**(1g)**

Fluorite (arrowed) in hydrothermally altered garnet-rich quartzo-feldspathic metavolcanic rock; Nisserdal region, Telemark, Norway.

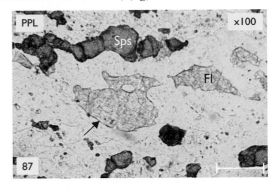

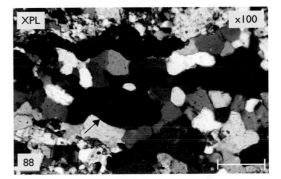

Fluorite (arrowed) in hydrothermally altered garnet-rich quartzo-feldspathic metavolcanic rock; Nisserdal region, Telemark, Norway.

Non-isotropic                                                        16

16   Relief varies on rotation ("twinkling"); high order birefringence.    17
     No change in relief on rotation; top 1st ord. birefr.; (lozenge-shaped) **(1h)**.

                                                        **LAWSONITE(□)**

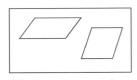

**(1h)**

Lawsonite (rectangular and lozenge-shaped sections) in glaucophane schist (blueschist); Orhaneli, NW Turkey (photo courtesy of Giles Droop).

17 Yell.-brn./orange-brown Fe-staining along clv. traces (1i)

**ANKERITE (or Fe-Calcite, Siderite)**

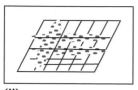

**(Ii)**

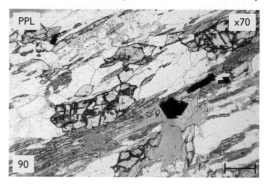

Ankerite (irregular porphyroblasts, showing 'rusty' alteration along cleavage) with quartz, chlorite and epidote in greenschist; Döllach, Austria (photo courtesy of Giles Droop).

No significant Fe-staining along cleavage traces                                          18

18 Some deformation twins (if present) bisect acute cleavage angle (1j)    **CALCITE**

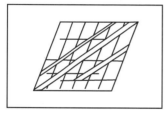

**(Ij)**

Calcite (with deformation twins) in marble; Troms, Norway.

Some deformation twins (if present) bisect obtuse cleavage angle (1k)

**DOLOMITE**

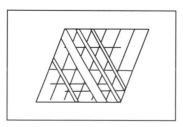

**(Ik)**

Dolomite in chl-talc-dol schist; Ghana

*(Note: Magnesite (of serpentinites) is optically indistinguishable from calcite/dolomite).*

**RHODOCHROSITE**

Rhodochrosite in vein; Sri Lanka.

*Note: Rarer RHODOCHROSITE (MnCO₃) (above) is also similar in appearance to calcite and dolomite, but is typically colourless to pale pink or pale pinkish-brown in PPL.*

19  **2 CLEAVAGES at c. 78–85° (≡ 102–95°)**                         **20**
    Cleavages at c. 90°                                             21

20  1st ord. yellow/orange, HIGH relief, 2V = 82 83°; regional met. metapelites, or
    some eclogites; common. (1l)                                 **KYANITE**

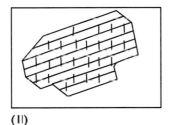

**(1l)**

Kyanite in ky-bt schist; Ross of Mull, Scotland.

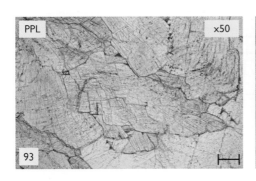

Max 1st ord. yellow/orange MOD (to high) relief, 2V = 37°; barium-rich rocks
and veins.                                                      **BARYTES**

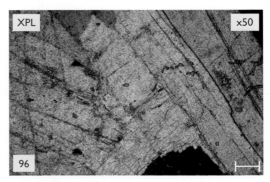

Barytes in vein; Porth y Corwgl, Anglesey, Wales.

*Note: Corundum with well developed parting could possibly be confused, see Section 5.*

21    **2 CLEAVAGES at 90° (Im or In);**

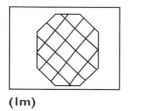

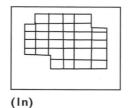

**(Im)**                    **(In)**

ISOTROPIC (or almost so)                                    22
ANISOTROPIC                                                 24

22    High relief (**1n**)                              **PERICLASE**
      Low or moderate relief                                 23

23    Colourless (primary mineral or vesicle fill of basic and intermediate igneous rocks)
                                                        **ANALCITE**

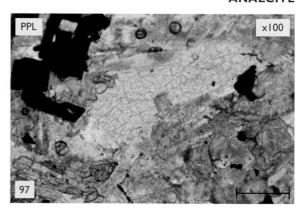

Analcite in teschenite; Edinburgh,
Scotland.

Colourless to pale blue; weak cleavage, but aligned inclusions common (primary
mineral in phonolites, nepheline syenites and other alkaline igneous rocks)

**SODALITE, NOSEAN, HAÜYNE**

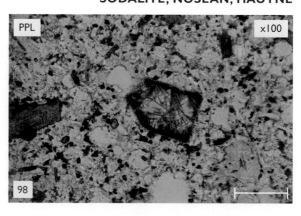

Haüyne    in    leucite    tephrite;
Tavolato, Rome, Italy.

24    Cross-hatched twinning evident (**1o**)                    **MICROCLINE**

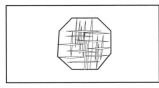

(**1o**)

Microcline in granite; Penmaenmawr, Wales.

*Note: Relatively rare ANORTHOCLASE is similar to microcline, but has ultra-fine twinning.*

Not showing cross-hatched twinning                                      25

25    Symmetrical (c.40–45°) extinction (**1p**)                          26

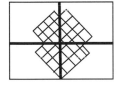

(**1p**)

Extinction // to cleavage (**1q**)                                      39

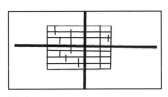

(**1q**)

26    1$^{st}$ order grey/white interference colours                      27
      Interference colours above 1$^{st}$ ord. grey/white                28

27    Mafic/ultramafic rock                                           **ENSTATITE(□)**

Enstatite (end and side sections) in boninite; Chichijima, Bonin Islands, Japan.

Meta-pelite/mudrock (Al-rich)                                    **ANDALUSITE(□)**

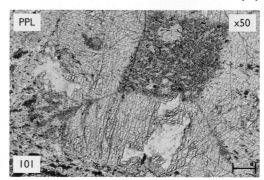

Andalusite (end-section) in crd-and slate; Connemara, Ireland.

28    Colourless                                                         29
      Weakly or strongly coloured in PPL                              32

29    Calc-silicate rock/marble/skarn                              **DIOPSIDE(□)**

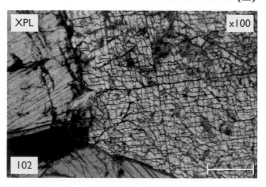

Diopside in calc-silicate skarn; Loch Ailsh, Assynt, Scotland.

Not calc-silicate rock/marble/skarn                                30

30    Blueschist facies metamorphic rock                          **JADEITE(□)**

Jadeite (end section) in blueschist;
Ile de Groix, Brittany, France.

*Note: Omphacite (colourless to pale grn. is similar).*

Igneous rock                                                              31

31    Granitic pegmatite (Li-rich)                           **SPODUMENE(□)**

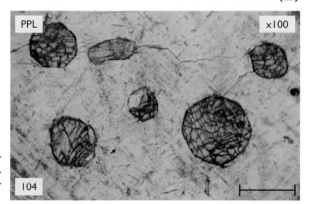

Spodumene (end-section) in granite pegmatite; Leinster, Ireland. (see start of Section 1 for enlargement of this image).

Alkali-olivine basalt & alkali igneous rocks.          **DIOPSIDE(□)**

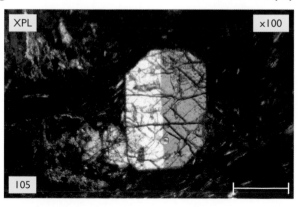

Diopside (end-section, with simple twin) in basalt; ODP Hole 793B, Izu-Bonin Arc, Japan.

32    Green/yell.-grn./violet or brown pleochroism.                          33
        Pleochroism weak or absent.                                           34

33    Green/yellow-green pleochroism              **AEGIRINE/AEGIRINE-AUGITE(□)**

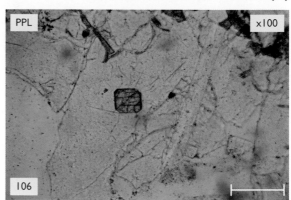

Aegirine-augite in leucite tephrite; Vesuvius, Italy.

Greenish/brownish to violet pleochroism              **TITANAUGITE(□)**

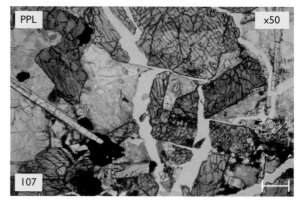

Titanaugite (end and side sections) in nepheline syenite; Saxony, Germany.

34    Weakly pleochroic                                                       35
        Non-pleochroic                                                        37

35   Max 1st ord. yell./orange interf. colours ($\delta = 0.010-0.016$). Weak pink to grn. pleochr.

**HYPERSTHENE($\square$)** (= $En_{50}Fs_{50} - En_{70}Fs_{30}$)

Hypersthene in granulite; Hartmannsdorf, Saxony, Germany.

1st ord. orange to 2nd ord. grn. interf. colours. Weak c'less to pale grn. or yell. grn. pleochr.                                                                                     36

36   2V 0–30°; weak c'less/pale yell.-grn. to brownish pleochr. 2nd ord. blue to bright green interf. colours ($\delta = 0.021-0.029$). Typical pyroxene of andesites and dacites.

**PIGEONITE($\square$)**

Pigeonite (end-section) in andesitic pitch-stone; Ardnamurchan, Scotland.

2V 56–84°; weak c'less to pale grn. pleochr. 1st ord. orange to 2nd ord. bright grn. iterf. colours ($\delta = 0.012-0.028$). Typical pyroxene of eclogites.

**OMPHACITE($\square$)**

An Fe-rich omphacite (end-section, with slight green to colourless pleochroism) with quartz and rutile in eclogite; Totaig, Glenelg, Scotland.   (photo courtesy of Giles Droop).

37    Calc-silicate rock/skarn (brownish grn.).                    **HEDENGERGITE(□)**

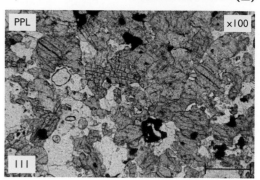

Hedenbergite (end and side sections) in skarn; Camas Malag, Skye, Scotland.

*(Note: v. similar to augite).*

Igneous rock (pale brown or greenish brown)                                    38

38    2V = 25–60° (biaxial +ve). Very common pyroxene in mafic and ultramafic igneous rocks.                                                        **AUGITE(□)**

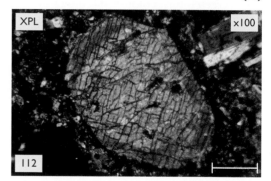

Augite (end section) in gabbro; Loch Coruisk, Skye, Scotland.

2V = 0–30°(biaxial +ve). Polysynthetic twins common in (100). Typical pyroxene of andesites and dacites.                                              **PIGEONITE(□)**

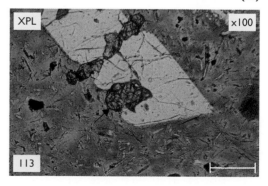

Pigeonite (end and side sections) in andesitic pitchstone; Ardnamurchan, Scotland.

39    1st ord. grey to v. pale yellow interference colours.                    **CELESTINE**

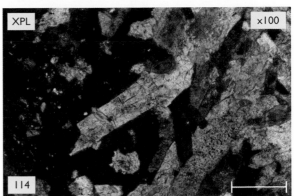

Celestine in Jurassic limestone;
Durlston, Dorset, England.

3rd order interference colours.                    **ANHYDRITE**

Anhydrite in anhydrite marl
from borehole; Selby, Yorkshire,
England.

# 1 Cleavage trace, inclined extinction

1   <u>1st ord. grey/white</u> (or pale yellow) interf. colours, with <u>multiple lamellar twins</u>
    (**2a**) or <u>streaky/braided</u> intergrowths (**2b**)or vermicular intergrowths (**2c**);
    colourless.                                                                                   2
    <u>Coloured</u>, <u>or</u> colourless, but lacking multiple lamellar twins or intergrowths (**2c**)
    (may show simple twins). If showing polysynthetic twins (**2a**) then higher than 1st
    ord. grey interference colours (**see 2m**)                                                   8

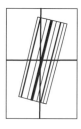      

**(2a)**                        **(2b)**                        **(2c)**

2   Streaky/braided intergowths            or        Vermicular intergrowths

   **PERTHITE (Microperthite)**              **MYRMEKITE** *(pl + qtz intergrown)*

(a) Microperthite in syenite; Fredricksvarn, Norway; (b) Myrmekite (i.e. plagioclase [mid grey
1st Ord. background] with intergrown vermicular quartz [white 1st Ord.]) in augen gneiss;
Bettyhill, Tongue, Scotland.

Multiple lamellar albite/albite-carlsbad twins (no braided intergrowths)
                                                                    **PLAGIOCLASE**   3

*Note: δ = 0.007–0.010 (1st order grey/white)*
*for Albite-Labradorite, but increasing to*
*δ = 0.010–0.013 (yellow) for Bytownite*
*and Anorthite.*

Plagioclase (labradorite) in troctolite; Sierra
Leone.

3   Max. Extinction angle < 12° (**2d**)                              **OLIGOCLASE**
    Max. Extinction angle >12°                                                    4

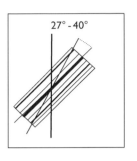

**(2d)**                    **(2e)**                    **(2f)**

4   Max. Extinction angle 12–27° (**2e**)                                         5
    Max. Extinction angle >27°                                                    6

5   Low +ve relief                                                    **ANDESINE**
    Low –ve relief                                                       **ALBITE**

6   Max. Extinction angle 27–40° (**2f**) ($\delta = 0.008-0.009$)   **LABRADORITE**
    Max. Extinction angle > 40° max. interf. colour to 1$^{st}$ ord pale yell.    7

7   Max. Extinction angle 40–52° (**2g**). ($\delta = 0.009-0.012$)  **BYTOWNITE**
    Max. Extinction angle 52–70° (**2h**). ($\delta = 0.012-0.013$)  **ANORTHITE**

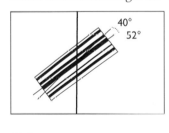

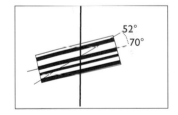

**(2g)**                                        **(2h)**

Plagioclase (bytownite) in anorthosite layer;
Rhum, Scotland. (see start of Section 2 for an
enlargement of this image).

8    Max. Extinction angle < 10° (2i) or (2j)                                    9
     Max. Extinction angle >10° (2k) or (2l)                                    16

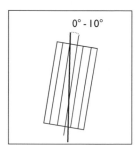

**(2i)**

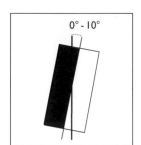

**(2j)**

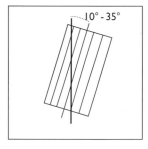

**(2k)**

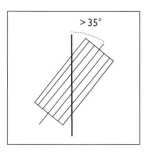

**(2l)**

9    Colourless                                                                  10
     Coloured                                                                    13

10   Max.1$^{st}$ ord. grey/white (often simple twins) (2j)                      11
     Higher that 1$^{st}$ ord.grey/white                                         12

11   2V <12°                                                     **SANIDINE**

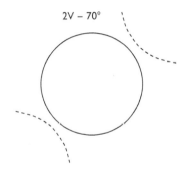

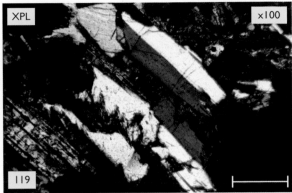

Sanidine in sanidinite facies mafic hornfels; Cushendell, Antrim, N.Ireland.

2V = 70°                                                         **ORTHOCLASE**

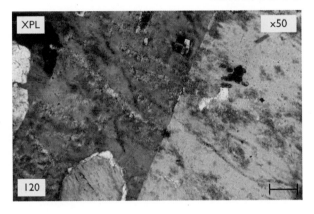

Orthoclase in granite; Carnmenellis, Cornwall, England.

*Note: Rhombic form of Orthoclase (ADULARIA) occurs in certain potassic alteration zones of hydro-thermal ore deposits.*

12   Max. 1$^{st}$ord. yell./orange interf. colours; common in high pressure metapelites and some eclogites. **KYANITE**

Kyanite in kyanite gneiss; Loch Assapol, Mull, Scotland.

Upper 3$^{rd}$ (to 4$^{th}$) ord. interf. colours; common accessory phase in granites, pegmatites and as detrital crystals in sedimentary/metasedimentary rocks. **MONAZITE**

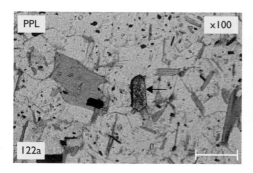

Monazite (arrowed) in Crd-Bt-Qtz-Ilm metasedimentary rock; locality unknown.

13   Grn. to grn.-brn., or yell-grn. to blue/blue-grn. pleochrosism.                14
     Colourless/blue/violet pleochroism                                            15

14   Green to green-brown pleochroism                          **AEGIRINE(▮)**

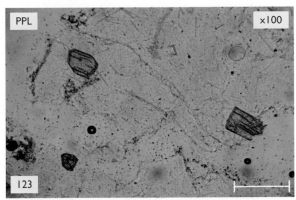

Aegirine-augite in leucite tephrite; Vesuvius, Italy.

Yell.-grn. to blue-grn./blue pleochroism.                   **ARFVEDSONITE(▮)**

Arfvedsonite in syenite; Oslo
region, Norway.

15    Colourless/blue pleochroism; max.ext. = 14°(often <10°)

**GLAUCOPHANE / CROSSITE(█)**

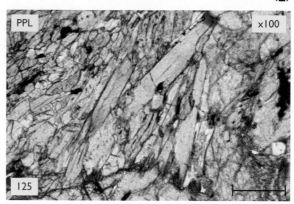

Glaucophane in glaucophane schist (blueschist); Ile de Groix, Brittany, France.

Intense prussian blue/indigo blue/green pleochroism.          **RIEBECKITE(█)**

Riebeckite in sodic granite; Drammen, Norway.

16    Max. Extinction angle 10–35° (**2k**)                                        17
      Max. Extinction angle >35° (**2l**)                                         30

17    Pleochroic                                                                  18
      Non-pleochroic                                                             24

18  Grn/yell., grn/blue-grn, grn/brn or yell.brn-red-brown pleochr.        19
    C'less/pale yellow pleochroism                                **CHONDRODITE**

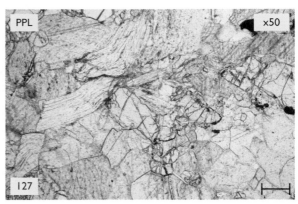

Chondrodite in mineralised skarn; Ottawa, Canada.

19  Max.extinction angle generally 10–19°                          20
    Max.extinction angle generally >19°                            23

20  Green, yellow-green, blue-green or greenish-brown pleochroism  21
    Yellow-brown, brown, orange-brown or red-brown pleochroism     22

21  Green to yell.-grn., or green to dk.grn./blue-green pleochroism. Greenschist
    metamorphic rocks, altered mafic rocks (common).     **ACTINOLITE(▮)**

*Note: Hornblende, ext. angle 12–34°, although usually with higher ext. angle, can be similar (see below). Pumpellyite, ext. angle 22° although usually very fine needles, is also similar to fine acicular actinolite.*

Actinolite in act-ab-czo hornfels; locality unknown.

Green/ yellow-grn. / greenish brown pleochroism; extinction angle 0–20°. Igneous rocks.                                                    **AEGIRINE-AUGITE(■)**

Aegirine-augite in leucite tephrite. The rim of the phenocryst and all the small coloured groundmass crystals are aegerine-augite composition. The core of the phenocryst has more augitic composition; Vesuvius, Italy.

Brownish yellow/pale yellow to red-brn./orange brn./ greenish brn. pleochroism. Extinction angle 0–19°. Igneous rocks.                      **KAERSUTITE(■)**

22   Strongly coloured; yell.-brn to red.brn or red.brn to grn-brn.    **KAERSUTITE(■)**

Note: *The amphibole KAERSUTITE, found in alkaline volcanic rocks, is very similar and thus difficult to distinguish from brown hornblendes (see below).*

Kaersutite phenocryst (side section - twinned) in syenogabbro ("lugarite"); Lugar Sill, Lugar, Ayrshire, Scotland.

Strongly coloured; pale yellow/yell.brn. to dark brn/chestnut brown or red.brn.
                                    **OXO-HORNBLENDE (brown Hornblende)(■)**

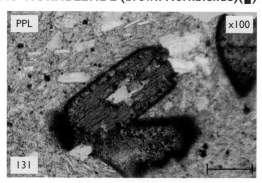

Note: *Crystals of kaersutite and oxo-hornblende are commonly seen with a rim of iron oxides (magnetite), formed during late stage oxidation (Deer et al., 2013).*

Oxo-hornblende (side section) phenocryst in andesite; Mt.Shasta, CA, USA.

23   Granular or radiating fibres (high rlf.); c'less to green, or pale yell.brn pleochr.; ext. angle = 4–22°; low temperature metamorphism.   **PUMPELLYITE**

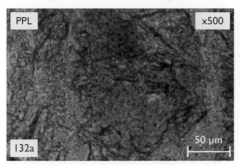

Pumpellyite (highly magnified) in sub-greenschist facies metavolcanic rock; Roseland, Cornwall, England. *Note: Acicular actinolite, ext. ang. 0–18°(see above), can look similar.*

Prismatic/rectangular (med/high rlf.); green/yell.-grn. /blue-green or brown pleochr.; max. extinction angle 13–34°(typically 20–34°); various igneous/ metamorphic rocks; (common)   **HORNBLENDE(▉)**

*Note: Most hornblendes are pleochroic in shades of yell.grn.-green-blue-grn. However, high temperature hornblendes may be green to brown, or brown pleochroic (Oxo-Hornblende). The amphibole KAERSUTITE (see above), found in alkaline volcanic rocks, with yell. brn.-red-brn.-greenish brown pleochroism and maximum extinction angle of 19° is difficult to distinguish from brown hornblendes.*

Hornblende in Hornblende gneiss; Sleat, Skye, Scotland.

24   1st order grey/white interference colours (ext.15–18°)
   **ZEOLITE mineral (e.g. SCOLECTITE)**
   Higher than 1st order grey interference colours   25

25    High relief, max. 1st order yellow/orange interf. colours.    **KYANITE**

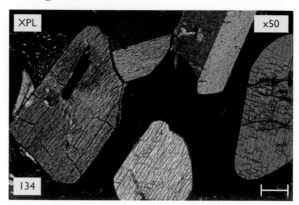

Kyanite in kyanite eclogite from kimberlite pipe; Roberts Victor Mine, South Africa.

Higher than 1st order orange interf. colours.    26

26    Top 1st order or 2nd order interference colours.    27
      3rd order birefr.; (ext.10–15°); often multiple lamellar twins; pale yellowy brown.
      **GRUNERITE( )**

Grunerite (side section) showing multiple-twinning in meta-iron-stone; Erikstad, Hinnøy, Lofoten Is., Norway (photo courtesy of Giles Droop).

27    Max. extinction 10–20°    28
      Max. extinction >20°    29

28  Abundant narrow polysynthetic twins // {100} (**2m**). Meta-ultramafics, other Mg-Fe rich rocks.  **CUMMINGTONITE(◼)**

**(2m)**

Cummingtonite (with characteristic polysynthetic twinning), in cum-grt-bt-qtz skarn/meta-ironstone; Zinkgruvan, Sweden.

Laterites, bauxites and other aluminous rocks.  **GIBBSITE(◼)**

*Note: In side-section* **GIBBSITE***, a common aluminium hydroxide of laterites and bauxites is similar, although crystals less elongate. Gibbsite often shows polysynthetic twinning parallel to {001}. Its birefringence is slightly lower than cummingtonite. In end-section the two minerals are easily distinguished as cummingtonite (an amphibole) shows two cleavages at 120°, whereas gibbsite has none, and appears as mica-/clay-like platelets.*

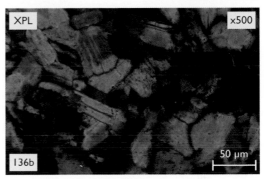

Gibbsite in laterite; Suriname.

Few or no twins. Calc-silicate (Ca-Si-Fe rich) rocks.  **TREMOLITE (◼)**

Tremolite in qtz-cal-tr calc-silicate rock; Mangkuma-Ketempe, Ghana.

29    Granitic pegmatites (max. extinction = 23–27°)                    **SPODUMENE(▮)**

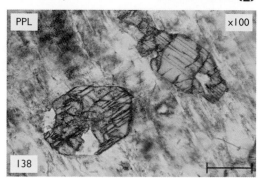

Spodumene (side section) in granite pegmatite; Leinster, Ireland.

Mafic igneous rocks (max. extinction = 22–45°)                    **PIGEONITE(▮)**

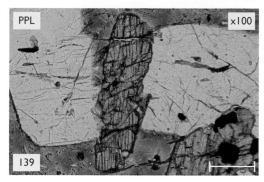

Pigeonite (side-section) in andesitic
pitchstone; Ardnamurchan, Scotland.

**Max. Ext. angle > 35° (generally 40–45°) (= Clinopyroxene)**

30    Colourless                                                                       31
      Coloured (including weakly coloured)                                          34

31    Max. top 1$^{st}$ ord. violet to low 2$^{nd}$ ord. bright blue interf. colours ($\delta = 0.006$–$0.021$).
      Blueschist facies metamorphic rock. Max.extinction 30–44°)        **JADEITE(▮)**

Jadeite (side section) in jadeite blueschist; Corsica, France.

Max. low to mid 2$^{nd}$ ord. interf. colours ($\delta = 0.023$–$0.031$). Not blueschist facies
metamorphic rock.                                                                   32

32   Calc-silicate rock (max.ext. 37–45°).                                       **DIOPSIDE(▌)**

Diopside in calc-silicate skarn;
Loch Ailsh, Assynt, Scotland.

Igneous rock                                                                              33

33   $2V = 0$–30° (generally colourless, but may show weak pleochroism to pale yel-
lowish/brownish-grn.)                                                         **PIGEONITE(▌)**

Pigeonite (side-section) in andesitic pitchstone; Ardnamurchan,
Scotland.

$2V = 58$–60°                                                                   **DIOPSIDE(▌)**

Diopside (twinned side-section)
phenocryst in basalt; Izu-Bonin
Arc, Japan.

34    Pleochroic (weak or strong)                                35
      Non-pleochroic                                             38

35    Strongly pleochroic                                        36
      Weakly pleochroic                                          37

36    Pleochroic green/brown to violet.              **TITANAUGITE(█)**

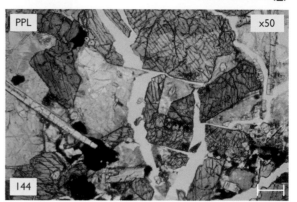

Titanaugite (end and side sections) in nepheline syenite;
Saxony, Germany.

Pleochroic from red-brn./dk. brn./brownish black              **AENIGMATITE(█)**

*Note: Also check brown
HORNBLENDE and
KAERSUTITE in the 10–35°
extinction angle part of the
key (couplet 16. pg.60).*

Aenigmatite (side section) showing cleavage and pleochroism
(from red-brown to virtually
opaque) in nepheline aenigmatite pegmatite; Kirovsk, Kola
Peninsula, Russia (photo courtesy
of Giles Droop).

37   $1^{st}$ ord. orange to $2^{nd}$ ord. bright grn. iterf. colours ($\delta = 0.012-0.028$). $2V = 56-84°$; weak c'less to pale grn. pleochr. Typical pyroxene of eclogites.   **OMPHACITE(▮)**

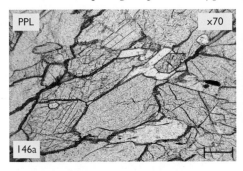
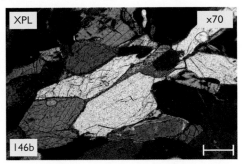

Omphacite (long- and cross-sections, with 'fuzzy' alteration rims) with phengite in eclogite; locality unknown (photos courtesy of Giles Droop).

$2^{nd}$ ord. blue to bright green interf. colours ($\delta = 0.021-0.029$). $2V$ $0-30°$; weak c'less/pale yell.-grn. to brownish pleochroism. Typical pyroxene of andesites and dacites.   **PIGEONITE(▮)**

Pigeonite (side-section) in andesitic pitchstone; Ardnamurchan, Scotland.

38   Calc-silicate rock/skarn (brownish green). ($\delta = 0.025-0.034$). $2V = 52-64°$.
   **HEDENBERGITE(▮)**

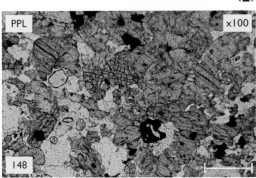

Hedenbergite (end and side sections) in skarn; Camas Malag, Skye, Scotland.

Not calc-silicate rock/skarn                                              39

39   Pale green. $1^{st}$ ord. orange to $2^{nd}$ ord. green iterf. colours ($\delta = 0.012–0.028$). 2V 56–84°. Typical pyroxene of eclogites.   **OMPHACITE(■)**

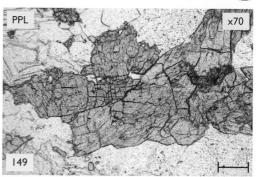

Omphacite (end and side sections) with phengite in eclogite; Syros, Greece (photo courtesy of Giles Droop).

Pale brown, yellowish brn., grn.brn., or yell.grn                              40

40   2V = 30–50°. ($\delta = 0.018–0.033$). Pale brown, yellowish brn. or grn.brn. Very common in mafic/ultramafic rocks).   **AUGITE(■)**

Augite (side section) in amygdaloidal basalt; Uig, Skye, Scotland.

2V = 0–30°. ($\delta = 0.021–0.029$). Pale yell.-grn. to brownish Typical pyroxene of andesites and dacites.   **PIGEONITE(■)**

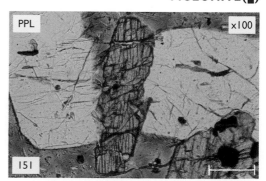

Pigeonite (side-section) in andesitic pitchstone; Ardnamurchan, Scotland.

# 1 Cleavage trace, straight extinction, colourless

1    High (or Mod./High) relief                                                      2
     Low/Moderate relief                                                             14

2    1st Ord. grey/white (or anomalous blue) interference colours.                    3
     Higher than 1st Ord. white interf. colours.                                      6

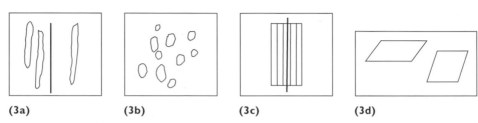

**(3a)**              **(3b)**              **(3c)**              **(3d)**

3    Fractured needles/rods (**3a**), or small granules (**3b**)                      4
     Rectangular or prismatic (**3c**), or rhombic (**3d**), but not rod-like.         5

4    RI 1.69–1.72 (high); $\delta$ = 0.003–0.008, but often anomalous blue.    **ZOISITE**

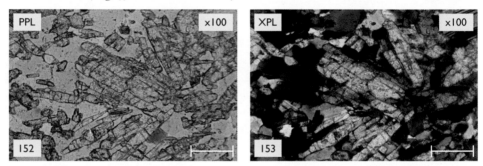

Zoisite in epidote amphibolite; Gratangen, Troms, Norway.

RI 1.61–1.64 (mod-high) $\delta$ = 0.008–0.010; accessory mineral in granitic rocks.

**TOPAZ**

*Note: Poorly formed small*
*ANDALUSITE crystals look similar.*

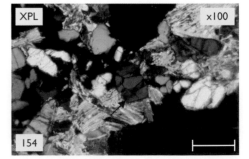

Topaz in greisen (i.e mica-rich hydrothermally altered rock) associated with granite;
St. Michael's Mount, Cornwall, England.

5   High relief, (biaxial +ve); ($\delta$ = 0.008–0.009). Crystals length-slow. Mafic/
ultramafic rock.                                                    **ENSTATITE(▮)**

Enstatite in boninite; Chichijima,
Bonin Islands, Japan.

Mod./high relief, (biaxial –ve); interf. colours reach 1st ord. yellow in some
crystals ($\delta$ = 0.009–0.012). Crystals length-fast. Metapelite/mudrock (Al-rich).
                                                                  **ANDALUSITE(▮)**

*Note: Also check Barytes, which is
much less common, but has
similar properties (see 8.
below).*

Andalusite (var. chiastolite) porphyroblasts in chiastolite slate;
Skiddaw Granite aureole, Lake
District, England. See Image 65
demonstrating length-fast nature
of this crystal

6   1st Order yellow/orange interference colours.                      7
     Higher than 1st order yell./orange interf. colours.                10

7   Mod-High relief (RI = 1.62–1.65).                                   8
     High relief (RI = 1.71–1.73).                                      9

8    Fibrous or rod-like aggregates (often radiating). Biaxial -ve, 2V = 36–60°. $\delta$ = 0.013–0.014. May show v.slight inclined extinction (0–4°). Calc-silicate rocks.    **WOLLASTONITE**

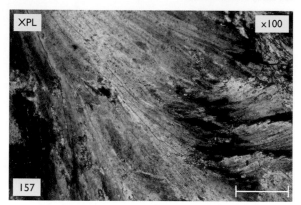

Wollastonite in calc-silicate rock in the contact aureole of the Dartmoor Granite; Meldon, Devon, England. Note the N-S oriented crystals (left) and E-W oriented crystals (centre right) showing straight extinction.

Prismatic/bladed crystals (sometimes radiating), or polygonal/granular aggregates. Biaxial +ve, 2V = 37°. Interf. colours range up to 1$^{st}$ ord. yellow ($\delta$ = 0.012), but most crystals are typically 1$^{st}$ ord. light grey/white. Occurs in Ba-mineralised areas/rocks.    **BARYTES**

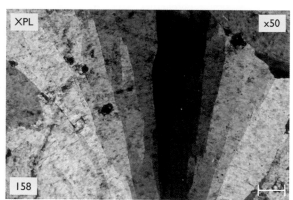

Barytes in vein; Porth y Corwgl, Anglesey, Wales. Note the N-S oriented central crystal showing straight extinction.

9   Rectangular/prismatic (**3c**), not rod-like; $\delta = 0.012–0.016$. $2V = 78–83°$. High-P metapelites and Al-rich eclogites.                                **KYANITE**

Kyanite in kyanite eclogite within kimberlite pipe; Roberts Victor mine, South Africa.

Needles, rods or small blebs (**3a, 3b**). $\delta = 0.004–0.015$.        **CLINOZOISITE**

Clinozoisite arrowed in albite of ab-act-chl-clz hornfels; Donegal, Ireland.

10  Top 1st order to mid 2nd Order interf. colours.                          11
    High order (i.e mid 3rd and 4th ord.) interf. colours.                   13

11  Squat rectangle/rhomb/lozenge (L/W typically = 2) (**3d**) Max. 1st ord. red/violet interf. colours ($\delta = 0.019–0.021$).                          **LAWSONITE**

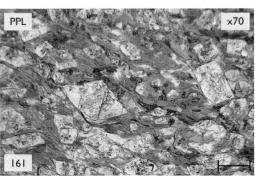

Lawsonite (rectangular and lozenge-shaped sections) in glaucophane schist (blueschist); Orhaneli, NW Turkey (photo courtesy of Giles Droop).

Elongate rectangle (L/W > 3) (**3e**). Max. 2nd ord. blue/grn. interf colours.    12

12    $\delta = 0.016-0.025$; biaxial +ve; 2V = 70–90°.                    **ANTHOPHYLLITE(█)**

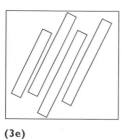

(3e)

Anthophyllite in cordierite-anthophyllite hornfels; Kenidjack, Cornwall, England. Note the N-S oriented crystal (upper centre) showing straight extinction.

$\delta = 0.020-0.033$; biaxial +ve; 2V 64–71°.                    **PREHNITE (█)**

Prehnite in Prh-qtz vein cutting volcaniclastic sediments; Roseland, Cornwall, England. Note the N-S oriented crystal (left) and E-W oriented crystal (right) showing straight extinction. Also see Fig. 1.12 for lower magnification view of the same vein.

13  Biaxial +ve (low 2V; 6–19°), small elongate blebs (**3b**); granitic rocks (rare accessory phase) and metasedimentary rocks; $\delta = 0.045$–$0.075$. Actually 2–10° extinction angle, but may appear parallel.    **MONAZITE**

Monazite in Crd-Bt-Qtz-Ilm metasedimentary rock; locality unknown.

Biaxial +ve (large 2V; 84–86°), platy, aluminous rocks (rare); $\delta = 0.040$–$0.050$.    **DIASPORE**

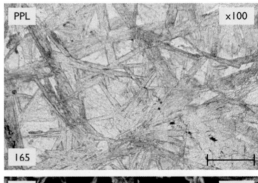

Diaspore; Wonderstone Quarry, South Africa.

14  1$^{st}$ order grey/white/yellow interference colours    15
    Higher than 1$^{st}$ order yellow interference colours    23

15    Fibrous (**3f**) or radiating acicular crystals (**3g**).                    16
      Anhedral/granular or prismatic crystals.                                     17

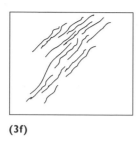

**(3f)**

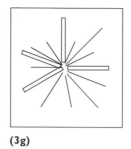

**(3g)**

16    Low relief

**ZEOLITE**
**(eg. THOMSONITE, MESOLITE)**
*colourless*

**or SERPENTINE**
**(e.g. ANTIGORITE)** usually c'less/pale green

Zeolites as cavity fill in altered gabbro; Scawt, Antrim, N.Ireland.

Serpentine aggregate in serpentinised ultra-mafic rock; Lizard, Cornwall, England.

Medium relief (1st order grey/white (sometimes to yellow) interf. colour)

**ZEOLITE (eg NATROLITE)**

Zeolite in cavity within leucitophyre; Eifel, Germany.

17   Fine-grained aggregate (3h)                    **CLAY MINERALS (e.g ILLITE)**

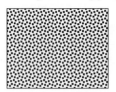

**(3h)**

Andalusite porphyroblast pseudomorphed to an aggregate of clay minerals and sericite in andalusite slate; contact aureole of Skiddaw Granite, Lake District, England.

Not fine-grained aggregate                                            18

18   Low relief                                                       19
     Medium relief                                                    21

19   Simple twinning (3i) typical, 1$^{st}$ ord. grey/white; $\delta = 0.006–0.008$. Some sections
     show // extinction, but others c.5–9°.                  **SANIDINE**

**(3i)**

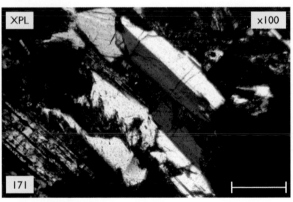

Sanidine in sanidinite facies mafic hornfels; Cushendell,
Antrim, N.Ireland.

Lacks simple twinning, v.low relief.                                  20

20    V.low relief. 1$^{st}$ ord. dk.grey/grey interf. colours; $\delta = 0.003-0.005$; igneous rocks.
**NEPHELINE**

Nepheline in nephelinite; Bohemia, Czech Republic.

Low -ve relief; 1$^{st}$ ord. grey/white interf. colours (occasionally to straw yellow), $\delta = 0.009$, sedimentary rocks (evaporites).    **GYPSUM**

Gypsum in evaporite; Kirkby Thore, Cumbria, England.

21    Med. -ve relief (= Zeolite mineral)    22
Med. +ve relief. Accessory mineral in granitic rocks.    **TOPAZ**

Cluster of topaz crystals in granite (Note: horizontal crystal in centre of image showing straight extinction); Cornwall, England.

22   Biaxial −ve figure                                                                **STILBITE**
     Biaxial +ve figure; typically associated with mafic metavolcanic rocks and basalts
     (low-T metamorphism).                                                            **HEULANDITE**

23   1$^{st}$ order yellow to mid 2$^{nd}$ order interf. colours.                      24
     Higher than mid 2$^{nd}$ order interf. colours.                                  28

24   Fine aggregates (**3h**)                                                          25
     Prismatic (**3c**), acicular (**3g**) or large anhedral crystals.                26

25   Max. 1$^{st}$ order violet interf. colours; $\delta = 0.012-0.020$. Calc-silicate rock/marble,
     or serpentinite.                                                                 **BRUCITE**

Brucite (showing upper-first-order interference colours) with calcite in brucite marble; Qadda, Saudi Arabia (photo courtesy of Giles Droop).

Interf. colours to mid 2$^{nd}$ order (and higher); typically $\delta = 0.010-0.036$.
certain **CLAY MINERALS (e.g. Smectite)**

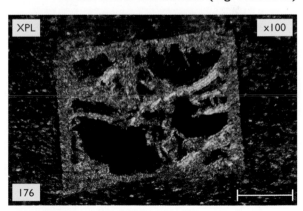

Clay minerals (e.g. smectite) with chlorite forming pseudomorph after andalusite in andalusite slate; Skiddaw granite aureole, Lake District, England.

26    Low relief
       Medium relief

27
**PREHNITE**

Prehnite in Prh-qtz vein cutting volcaniclastic sediments; Roseland, Cornwall, England. Also see Images 25 and 163 for XPL view of the same vein.

Prehnite as cavity fill in prehnite-pumpellyite facies metabasalt; Barrhead, Glasgow, Scotland.

**or HEMIMORPHITE**

Hemimorphite (Zn-silicate) found associated with Zn-ore deposits is optically similar to prehnite, although often acicular. This example is in a Zn-mineralised siliceous rock; Minera, Wrexham, Wales. Note the N-S and E-W oriented crystals in the radiating aggregates displaying straight extinction. Barytes and celestite, both at the high end of medium relief may also appear similar, as can certain medium relief zeolites that have $1^{st}$ order yellow/orange birefringence (e.g. natrolite $\delta = 0.012$).

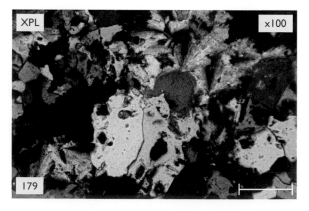

27   Low (+ve) relief; calc-silicate rock/skarn.                    **SCAPOLITE**

Scapolite in skarn; Arendal, Norway.

Low (-ve) relief; nepheline syenite or related rock.            **CANCRINITE**

Cancrinite in understaurated kfs-feldspathoid alkali igneous rock termed miascite; Miass, Urals, Russia. Note the straight extinction of the E-W oriented crystal left of centre.

28   Mod. relief. Elongate rectangles (**3e**), laths (**3c**), or fine aggregates (**3h**).   29
     Low (or mod.) relief; rectangular laths (**3c**) or short squat prisms (**3d**).        31

29   $2V = 0-30°$. Marble, calc-silicate or meta-ultramafic rock.    **TALC**

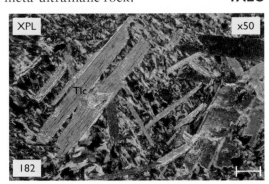

Talc in chl-talc-dol schist; Ghana. (see start of Section 3 for an enlargement of this image).

$2V > 30°$. Not marble, calc-silicate or meta-ultramafic rock.    30

30    Radial aggregates of crystals (**3g**) or "bow-tie" (**3j**). 2V= 53–62°. Aluminous pelites and hydrothermal ore deposits.                    **PYROPHYLLITE**

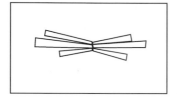

**(3j)**

Pyrophyllite in pyrophyllite rock; Tres Cerritos, Mariposa County, California, USA.

Not radiating aggregates. 2V = 28–47°. Very common mineral in a wide range of rock types.                    **MUSCOVITE**

Muscovite in muscovite schist; Ghana.

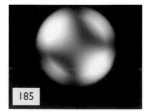

Muscovite interference figure, with isogyres arching half-way across the field of view with a boomerang-like angular curvature indicative of 2V c.30–45°.

Notes: (a) Fine grained muscovite (or paragonite) is referred to as SERICITE. (b) **PARAGONITE** (Na-white mica) is much rarer than MUSCOVITE, but optically indistinguishable. (c) LEPIDOLITE, a lithium mineral of certain granite pegmatites is also very similar; see 31. below).

31   Low to Mod relief. RI = 1.52–1.59. Max. top $2^{nd}$ ord. interf. colours ($\delta = 0.018$–0.038). Biaxial −ve figure; 2V = 0–58°; granite pegmatites and high T veins. Crystals may show very slightly inclined extinction (up to 6°).

**LEPIDOLITE or ZINNWALDITE**

Lepidolite (colourless in PPL) in aplite; Meldon Quarry, Devon, England.

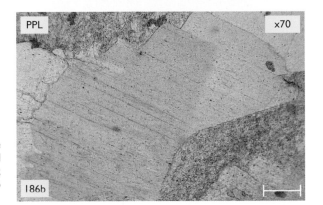

Zinnwaldite (colourless to v.pale brown in PPL) with altered feldspar in zinnwaldite granite; Gunheath, Cornwall, UK (photo courtesy of Giles Droop).

Mod. rlf. RI = 1.57–1.61. Low to mid 3rd ord. interf. colours ($\delta = 0.040$); Biaxial +ve figure; sedimentary rocks, especially associated with salt deposits and evaporites.                                                              **ANHYDRITE**

Anhydrite in anhydrite marl from borehole; Selby, Yorkshire, England.

# 1 Cleavage trace, straight extinction, coloured

1   Non-pleochroic                                                    2
    Pleochroic                                                        3

2   Red-brown (v.high relief)                                    **RUTILE**

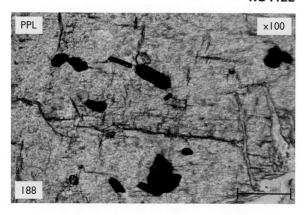

Rutile inclusions in kyanite, within kyanite schist; Loch Assapol, Mull, Scotland. Note: Much rarer, cassiterite (Sn-ore mineral) is also a very high relief brown or yellow-brown (or red-brown) mineral with high interference colours. It has polygonal to short prismatic form. (see pg. 107 (photo 231) and pg. 137 (photo 296a–b)

Colourless-very pale blue (high relief)          **DIASPORE**

Diaspore; Wonderstone Quarry, South Africa.

3   Weak/mod. pleochroism (pink/pale brn./pale grn.).          4
    Strong pleochroism.                                        5

4   Weak pleochroism; 1st order yellow/orange interf. colours   **HYPERSTHENE(▮)**

**(= orthopyroxene)**

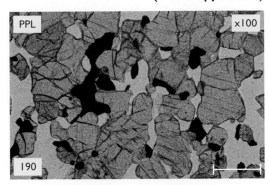

Hypersthene in granulite; Hartmanns-dorf, Saxony, Germany.

Mod. pleochroism: Top 1st/Low 2nd ord. interf. colours.   **FERROSILITE(▮)**

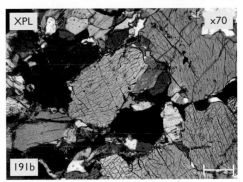

Ferrosilite (in various orientations, showing cleavage and pleochroism) with quartz in garnet-quartz-ferrosilite-granofels (a meta-ironstone); Mt Towla area, Zimbabwe (photos courtesy of Giles Droop).

5   Red to violet to orange-yell. pleochroism (depends on section)   **PIEMONTITE**

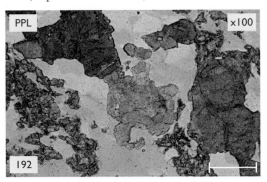

Piemontite. Pennsylvania, USA.

Not extreme red/violet/orange/yellow pleochroism.                                    6

6    C'less/pale yell./bright yell./yell.-green pleochr.                          7
     Pleochr. in green, brown (includes yell. → brown), or blue/violet.           8

7    1$^{st}$ order yellow interf. colours (colourless/pale yell./yell. pleochroism). Commonly
     as porphyroblasts. Weak cleavage trace.                          **STAUROLITE**

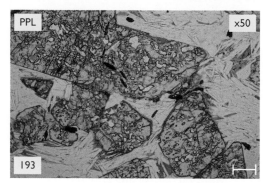

Staurolite porphyroblasts in stt-bt-schist;
Cloncurry, Queensland, Australia.

Max. high 2$^{nd}$ ord interf.colours (c'less/pale yell./yell.-grn. pleochr.). High relief
granular aggregates/ small rods.                          **EPIDOTE**

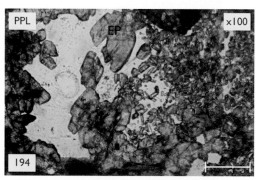

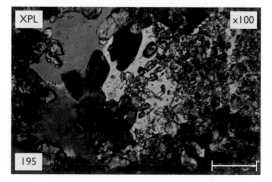

Epidote in epidote-actinolite hornfels;
Ghana.

8    Blue or blue/violet pleochroism.                          9
     Green, blue-grn., or green/brown pleochroism.                          11

9   Granite, syenite or aluminous metasediment.                              10
      Not granite, syenite or Al-metased (c'less/blue-violet).   **GLAUCOPHANE(▮)**

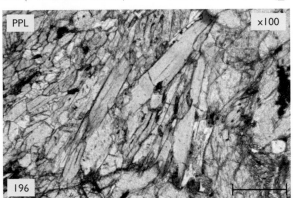

Glaucophane in glaucophane schist (blueschist); Ile de Groix, Brittany, France.

10   Blue/light blue/blue-grn. plchr. (Na-rich igneous rock).   **RIEBECKITE(▮)**

Riebeckite in sodic granite; Drammen, Norway.

C'less/blue pleochr. Al-metased or granite pegmatite.   **DUMORTIERITE**

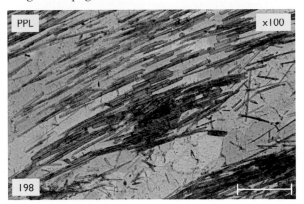

Dumortierite in dumortierite-qtz rock; locality unknown. (see start of Section 4 for an enlargement of this image)

11   C'less to **GREEN** (or blue-grn./tourquoise) pleochr.                    12
      C'less/pale **BROWN** to dk.brn., yell-brn, or brn.- grn. pleochr.        15

12  High relief                                                                    13
    Medium (or low) relief                                                         14

13  Prismatic/rectangular **(4a)**, often with "hour-glass" structure **(4b)**, or lamellar
    twinning.                                                        **CHLORITOID**

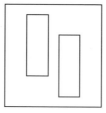

**(4a)**

Chloritoid in chloritoid phyllite; Tarkwa,
Ghana.

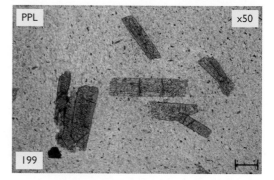

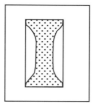

**(4b)**

Chloritoid [*ottrelite*] (with hour-glass
zoning) in chloritoid phyllite; Ardennes,
Belgium.

Acicular/rodded **(4c)** or columnar. Cross-fractures evident in rods, but cleavage
traces are weak.                                                **TOURMALINE** (schorl)

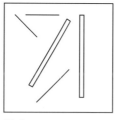

**(4c)**

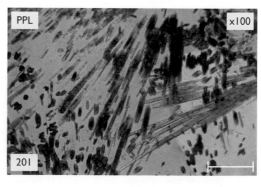

Radiating acicular tourmaline (schorlite) in Luxullyanite; Cornwall, England. Note the strong
pleochroic scheme as displayed by differently oriented crystals in top right quadrant of image.

14 C'less to pale grn., 2nd/3rd order birefringence. **PHENGITE** (+ some Biotites).

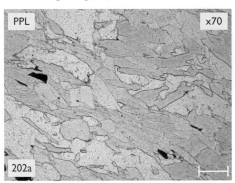

Phengite in quartz phengite schist; Mallnitz, Austria (photo courtesy of Giles Droop).
*Note: Cr-muscovite (FUCHSITE) is similar.*

C'less to pale grn., 1st ord. grey or anomalous "berlin blue".          **CHLORITE**

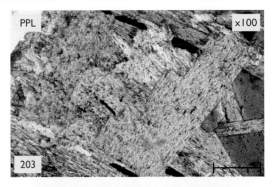

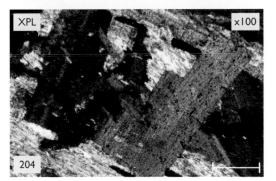

Late overprinting chlorite porphyroblasts in garnet mica schist;
Gratangenfjord, Troms, Norway. Note
the anomalous 'Berlin blue' interference
colours in XPL.

15 Medium relief                                                          16
   High relief                                                            18

16  Mottled (**4d**) at extinction                                                17

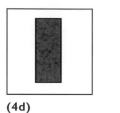     

**(4d)**                              **(4e)**

Smooth (**4e**) at extinction (intense yell.-brn. pleochr.)          **STILPNOMELANE**

Stilpnomelane in blueschist facies sili-
ceous    meta-ironstone;    Laytonville
Quarry, CA, USA. Note the strength
of pleochroism as shown by the central
N-S oriented crystal compared with the
E-W oriented crystals top and bottom.

17  Strongly pleochroic (pale brn. or yell.-brn. to dark brn./red-brn).    **BIOTITE**

          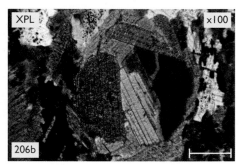

Biotite in grt-crd-sil schist; La Hoya, Andalucia, Spain.

Pale yell./pale brn. pleochr. (in marble, ultrabasic rock).          **PHLOGOPITE**

Phlogopite;    Pfaffenreuth,    Bavaria,
Germany.

18 Light brown-brown pleochroism. 1$^{st}$ order orange $\delta = 0.013$ to 2$^{nd}$ order red
   $\delta = 0.036$ interference colours.                                                    **ALLANITE**

Allanite in vein associated with
diorite; West Labelle Diorite,
Great Slave Lake, Canada.

Brown to yellow pleochroism. 4$^{th}$ order interference colours, $\delta = 0.060$ but often
masked by strong absorption colours.                                              **ASTROPHYLLITE**

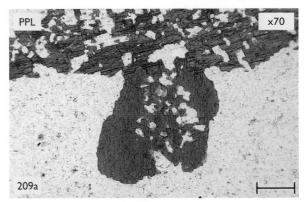

(209a) Astrophyllite (showing cleavage and pleochroism)
in syenite; Greenland (photo
courtesy of Giles Droop).

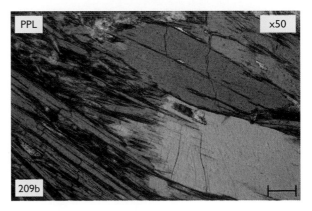

(209b) Astrophyllite in syenite;
Kola Peninsular, Russia.

# Imperfect cleavage, parting or arranged fractures

1    Relief varies on rotation ("twinkling") (5a); extreme 5th order interf. colours.

**Carbonate mineral: e.g. CALCITE, DOLOMITE or ARAGONITE.**

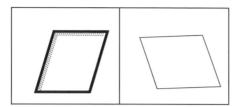

Note: Rarer RHODOCHROSITE ($MnCO_3$) is also similar in appearance, but is typically colour-less to pale pink in thin section.

**(5a)**

*CALCITE*

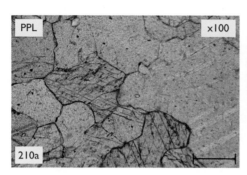
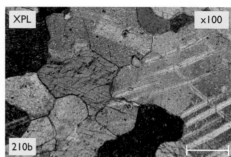

Calcite in marble; Troms, Norway.

*RHODOCHROSITE*

Rhodochrosite in vein; Sri Lanka.

Relief stays constant on rotation; generally 1st to 4th order interef. colours.    2

2    High relief                                                                    3
     Low or moderate relief                                                         20

3    Pleochroic                                                                     4
     Non-pleochroic                                                                 10

4    C'less to yell. or pale yell. to bright yell., or pale yell. to orange-brn.                    5
     Pleochr. is green, blue-grn., brown or purple.                                                  8

5    1st ord. yell./orange interf. colours ($\delta$ = 0.011-0.014); metapelites (often porphy-
     roblasts). Colourless to yellow pleochroism.                              **STAUROLITE**

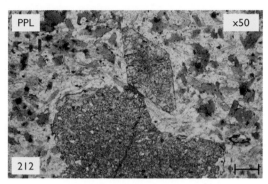

Staurolite in stt-bt-schist; Nauyago, Ghana.

     Higher than 1st ord. yell./orange interf. colours.                                             6

6    2nd order interf. colours.                                                                      7
     3rd order bordering 4th ord. interf. colours (yellow-peach to pale pink)
     ($\delta$ = 0.052). C'less to pale yellow pleochroism.                        **FAYALITE**

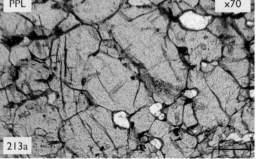

Fayalite in various orientations, showing weak pleochroism in pale yellows
(213a), with garnet, (isotropic) apatite
and rare hedenbergite (213b) in euly-
site (a meta-ironstone); Tunaberg,
Södermanland, Sweden (photos courtesy of Giles Droop).

7    Max. low to mid $2^{nd}$ ord. interf. colours (bright blue to bright green) ($\delta$ = 0.021-0.029). C'less to yellow pleochroism. Uniaxial +ve interf. figure.

**TOURMALINE (Dravite)**

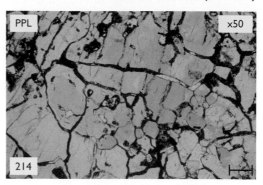

Tourmaline (Dravite); locality unknown.

Mid to upper $2^{nd}$ ord. interf. colours ($\delta$ = 0.028-0.037). C'less to pale yellow or golden yellow (sometimes yell.-grn.) pleochrosim. Biaxial +ve interf. figure. Calc-silicate rocks (skarns).    **HUMITE Gp.   (e.g Chondrodite, Clinohumite)**

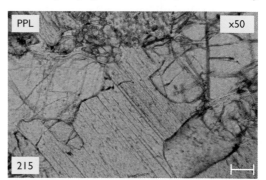

Clinohumite in mineralised skarn; Ottawa, Canada.

8    Spectacular purple to brown pleochroism.    **YODERITE**

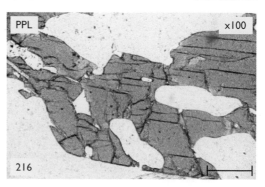

Yoderite in yoderite-tlc-qtz schist; Mpwapwa, Tanzania.

Not purple to brown pleochroism.    9

9   Green to blue-grn. or grn. to yell.-grn. pleochr.        **TOURMALINE (Schorlite)**

Tourmaline (end and side sections) in quartz-tourmaline rock; Roche, Cornwall, England.

Pale brown to brown pleochroism:                    **ALLANITE** (= *ORTHITE*)

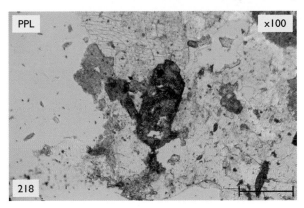

Allanite in syenite; Ghana.

| | | |
|---|---|---|
| 10 | Colourless | 11 |
| | Brown, yellow or yell.-brn; v.high relief. | 19 |
| 11 | 1$^{st}$ order to low 2$^{nd}$ order interference colours | 12 |
| | Mid 2$^{nd}$ order interference colours or higher | 18 |
| 12 | 1$^{st}$ order grey/white interf. colours (and/or anomalous purplish/blue). | 13 |
| | Higher than 1$^{st}$ order grey/white interf. colours | 15 |

13   Small squares (b) often clustered together (schist, pelitic hornfels). 1st order grey.
**SILLIMANITE(■)** *(or poorly formed ANDALUSITE)*

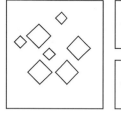

Sillimanite

Andalusite

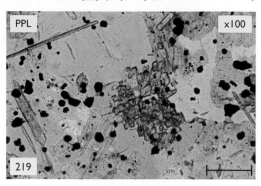

**(5b)**

Sillimanite (end-sections) in sil-bt-hornfels; Aberdeenshire, Scotland.

Polygonal, rectangular or anhedral crystals. 1st order grey/white or anomalous berlin blue interf. colours. Often zoned.                                            14

14   Polygonal or anhedral crystals, often with concentric zoning (5c). High relief (RI= 1.70-1.75) (calc-silicate/skarn):        **VESUVIANITE (=IDOCRASE)**

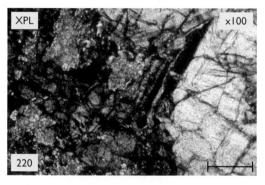

**(5c)**

Vesuvianite (=idocrase) in calc-silicate skarn of contact metamorphism; Loch Auvre, Donegal, Ireland.

*Note: Ca-garnets (e.g. grossular) are often anisotropic and show similar concentric zoning.*

Rectangular crystals with straight extinction, and commonly with a central dark band/line in XPL. Mod/High relief (RI = 1.62-1.66).        **MELILITE**

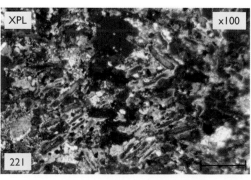

Melilite in vananzite (an undersaturated igneous rock); San Venanzo, Italy.

15    1st order straw yellow interf. colours. May show weak c'less to pale mauve/violet pleochr. in thick sections. (Contact metamorphic calc-silicate rocks and granite pegmatites).                                                                                        **AXINITE**

Axinite in altered dolerite near granite intrusion; St.Columb Road, Cornwall, England.

Max. upper 1st order or low 2nd order interference colours.                                    16

16    Rodded crystals with cross-fractures giving a "bamboo-like" appearance (**5d**) ($\delta$ = 0.018-0.022).                                                                                    **SILLIMANITE**

(**5d**)

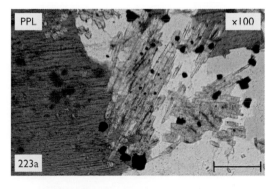

Sillimanite rods in sil-bt-hornfels (also note cluster of square end-sections top left); Weets, Grampian, Scotland.

Anhedral granular crystals or subhedral prismatic crystals.                                    17

17    1 imperfect parting (max.1st order red interf. colours).    **MONTICELLITE**

Monticellite in monticellite-magnetite skarn; Camas Malag, Skye, Scotland.

2 parting traces at 75-80° (e) (twinning common).    **CORUNDUM**

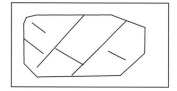

**(5e)**

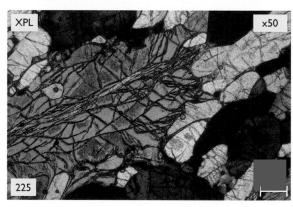

Corundum (with alteration to margarite along cracks) in anyolite (zoisite-edenitic hornblende-corundum(ruby) rock); Arusha region, Tanzania (see start of Section 5 for an enlargement of this image).

*(Note:  Corundum should be mid 1st order birefr. ($\delta = 0.008-0.009$) for 30μm thin-section, but extreme mineral hardness means that it is often thick, and thus gives interf. colours to low 2nd order)*

18    Max. upper 2$^{nd}$ ord. interf. colours:                    **OLIVINE (Forsterite)**

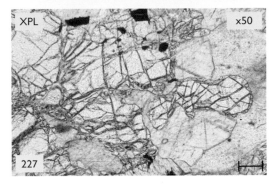

Olivine (forsterite) in contact metamorphosed olivine marble; Ottawa, Canada.

3$^{rd}$/4$^{th}$ order interf. colours (in granite pegmatites and as detrital grains in sedimentary and metasedimentary rocks).                    **MONAZITE**

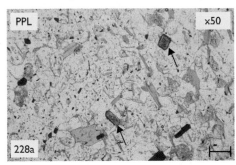

Monazite (arrowed) in Crd-Bt-Qtz-Ilm metasedimentary rock; locality unknown.

19   Isotropic                                                           **SPHALERITE**

Sphalerite in Zn-mineralised quartz-rich host rock; Zinkgruvan, Sweden.

*Note: Certain coloured garnets (e.g. spessartine) are also yellow-brown and isotropic, but don't really have arranged fractures (see Section 7).*

High ord. interf colours; diamond-shape (**5f**), or granular (**5g**).

                                                                  **TITANITE (=Sphene)**

**(5f)**                    **(5g)**

(a)  Titanite (=sphene) in augen gneiss; Bettyhill, Tongue, Scotland.

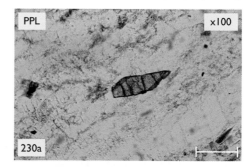

(b) Euhedral (diamond-shaped) titanite in granite; locality unknown.

*Note:  Much rarer, cassiterite (Sn-ore mineral) is also a very high relief brown mineral with high inter-ference colours. It has polygonal to short prismatic form (see image 231, top of next page, and images 296a-b).*

**CASSITERITE**

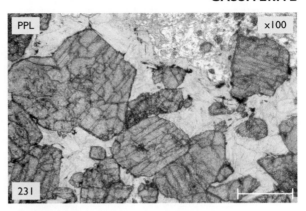

Cassiterite in cassiterite-rich vein rock within granite; East Pool and Agar Mine, Cornwall, England (also see image 296a–b).

20    Moderate relief.                                                        21
    Low relief (1st order grey/white interf. colours).        **NEPHELINE**

Nepheline in nephelinite; Bohemia, Czech Republic.

21    Yellow pleochroic (c'less to pale yellow, golden yellow, yell./brn. or orange-yellow). Towards the high end of moderate relief.

**HUMITE Gp. (e.g. Chondrodite, Clinohumite)**

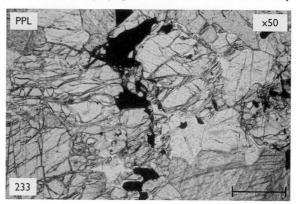

Chondrodite in mineralised skarn; Ottawa, Canada

Non-pleochroic                                                                                  22

22    Isotropic (colourless or pale blue).    **SODALITE/ NOSEAN/ HAÜYNE**

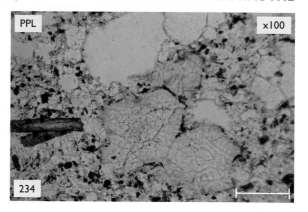

Haüyne in leucite tephrite; Tavolato, Rome, Italy.

1st order grey (or anomalous Berlin blue interf. colours)                      23

23    Straight extinction, commonly with a central dark band/line in XPL (often shows anomalous Berlin blue interference colours).    **MELILITE**

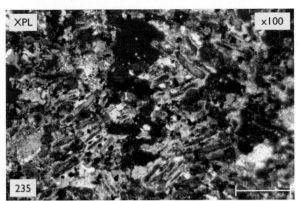

Melilite in vananzite (an under-saturated igneous rock); San Venanzo, Italy.

Symmetrical extinction (relative to clv./parting traces).    **CHABAZITE**

(236a) Chabazite-filled amygdale within basalt; Réunion Island (photo courtesy of Giles Droop).

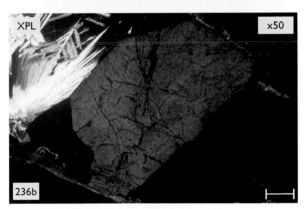

(236b) Chabazite showing perfect rhombohedral form due to unimpeded growth from vesicle wall; Talisker Bay, Skye, Scotland.

# 0 Cleavage traces, colourless

1    Relief varies on rotation (very high order interf. colours, granular).

**CALCITE, DOLOMITE or ARAGONITE (...)**

Note: microcrystalline calcite (grains <5μm), of sedimentary rocks is termed MICRITE.

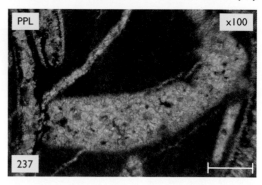

Granular calcite (micrite) in biomicritic limestone; Purbeck, Dorset, England.

Relief stays constant on rotation.                    2

2    **HIGH RELIEF**                    3
Low or moderate relief                    14

3    **Isotropic:**                    **GARNET**
[Almandine (Fe), Pyrope (Mg), Spessartine (Mn), Grossular (Ca)]

Note: Although largely colourless in thin-section, some garnets may show colouration (see Section 7). Almandine (pale pink); Ti-andradites (pale yell-brn.-brn.), Uvarovite (green). Note: Grossular commonly shows a degree of anisotropy, with zoning picked out by variations in 1st order grey interf. colours.

Garnet (almandine-rich) in regional metamorphic garnet mica-schist; Garve, Scotland.

Anisotropic                    4

4    1st order grey/white (or anomalous Berlin blue interf. colours).          5
     Higher than 1st order grey interf. colours.                              7


**1st order grey/white (or anomalous Berlin blue) interference colours.**

5    Biaxial +ve, // extinction, often shows anomalous Berlin blue interf.    **ZOISITE**

Zoisite in qtz-zo-grt-hbl hornfels; Foxdale, Isle of Man.


     Uniaxial –ve interference figure, // extinction.                         6

6    Often shows chemical (optical) zonation (see **5c**). Calc-silicate rocks.
                                                    **VESUVIANITE (= Idocrase)**

*Note: Grossular and other Ca-garnets can be anisotropic with low 1st ord. birefringence, and thus difficult to distinguish from vesuvianite.*

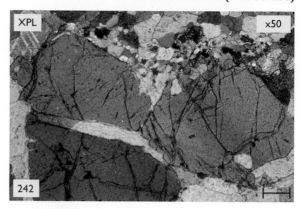

Vesuvianite (=idocrase) in marble; Loch Auvre, Donegal, Ireland.

Not chemically zoned. May show lamellar twinning. Al-rich, Si-poor rocks. Mostly uniaxial –ve, but some figures are biaxial 0–30°.    **CORUNDUM**

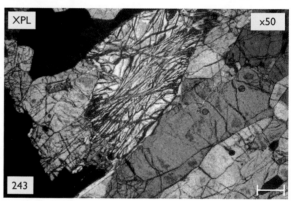

Corundum in anyolite (zoisite-edenitic hornblende-corundum(ruby) rock); Arusha region, Tanzania.

7    1st order yellow-orange interference colours.                          8
     Higher than 1st order yell./orange interf. colours.                    11

**1st order yellow/orange interference colours**

8    Pelitic rock or Al-rich silica-poor rock.                                    9
     Calc-silicate rock or mafic rock.                                            10

9    Fibrous or hair-like mat of crystals (6a), // extinction (metapelite)

**FIBROLITE (fibrous SILLIMANITE)**

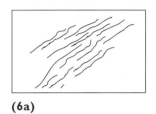

**(6a)**

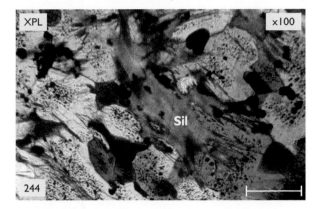

Sillimanite (fibrolite) in sillimanite
schist; La Hoya, Spain.

Granular (6b) or porphyroblastic, // extinction, very high relief.    **CORUNDUM**

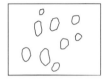

**(6b)**

Corundum in anyolite (zoisite-
edenitic hornblende-corundum
(ruby) rock); Arusha region,
Tanzania.

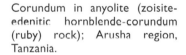

10   Biaxial –ve, // ext. (most sections); high-T calc-silicate rocks/skarns. Med to high
relief.                                                                    **WOLLASTONITE**

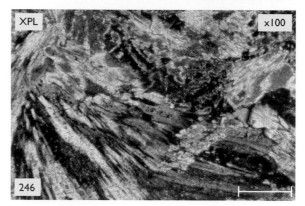

Wollastonite in calc-silicate rock
in the contact aureole of the
Dartmoor Granite; Meldon,
Devon, England.

Biaxial +ve, // ext. (length sections), calc-silicate and mafic rocks: **CLINOZOISITE**

Clinozoisite in albite of ab-act-
chl-clz hornfels; Donegal, Ireland.

## Top 1st order and low 2nd order interference colours

11  Top 1st order and low second order interf. colours.                    12
    Mid second order or higher interf. colours.                           13

12  Acicular or rodded ("bamboo cane") appearance, // extinction.   **SILLIMANITE**

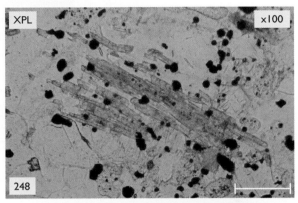

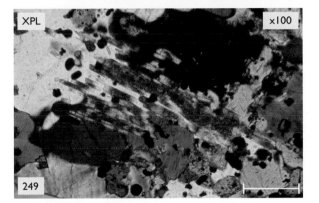

Sillimanite showing rodded "bamboo-like" appearance in kfs-sil-bt hornfels; Weets, Grampian, Scotland.

Small granules (**6b**), // extinction, very high relief.          **CORUNDUM**

*Note: Corundum should be mid 1st order birefr. for 30μm thin-section, but extreme mineral hardness means that it is often thick, and thus gives interference colours to low 2nd order.*

## Max. interference colours mid 2<sup>nd</sup> order or higher

13    Mid to upper 2<sup>nd</sup> order interference colours:    **OLIVINE (Forsterite)**

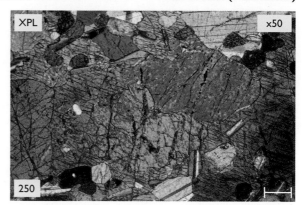

Olivine (forsterite) in gabbro cumulate; Duluth, Minnesota, USA.

3<sup>rd</sup>/4<sup>th</sup> order interf. colours:    **ZIRCON/ MONAZITE**

Zircon inclusion with well developed pleochroic halo in biotite of garnet mica-schist. Although this is an exceptionally well-formed and relatively large zircon, small zircon grains with dark pleochroic haloes giving a "mouldy" appearance to the phyllosilicate mineral are commonplace in biotites and chlorites (also see Image 313a). The haloes are considered to result from decay of radioactive elements within the zircon; Gratangenfjord, Troms, Norway.

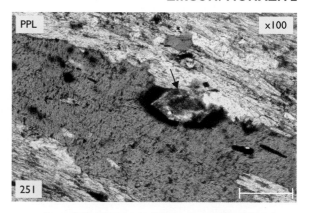

Zircon inclusion in biotite showing 3<sup>rd</sup> order interference colours and a feint pleochroic halo in the surrounding biotite; kyanite gneiss from Loch Assapol, Mull, Scotland.

## LOW OR MODERATE RELIEF

14  Moderate relief                                                            15
    Low relief                                                                 26

## MODERATE RELIEF

15  Isotropic or v.dark to mid grey 1$^{st}$ order interf. colours (close to isotropic);
    $\delta = 0.000 - 0.004$.                                                  16
    1$^{st}$ order mid grey/white, or higher order birefringence               20

16  Fibrous or acicular form (6a):                    **ZEOLITE Gp. (e.g. MESOLITE)**

*Note: Mod. −ve relief, bordering on low −ve relief; RI = 1.505–1.512; δ = 0.001.*

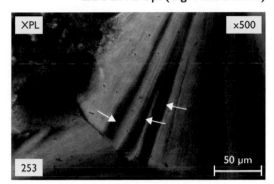

Mesolite needles (1$^{st}$ order very dark grey $\delta = 0.001$), interspersed amongst radiating needles of thomsonite/natrolite (1$^{st}$ order light grey to white) as part of vesicle infill, in amygdaloidal basalt; Spray, Oregon, USA.

Not fibrous/acicular                                                           17

17  Complex cross-hatched twinning (6c). Hexagonal, or trapezohedral shape;
    RI = 1.508–1.511, therefore bordering low/mod −ve relief; $\delta = 0.001$.    **LEUCITE**

(6c)

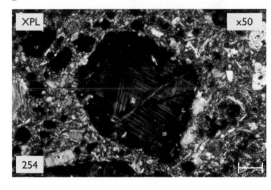

Leucite in leucite tephrite; Vesuvius, Italy.

Lacks cross-hatched twinning. May be rounded, hexagonal, trapezohedral or short rods/prisms.                                                             18

18    "Bright" relative to other colourless minerals; hexagon (6d)/granular/short rod/
prism). Med. to high relief (RI = 1.63–1.655); $\delta$ = 0.001–0.007.    **APATITE**

**(6d)**

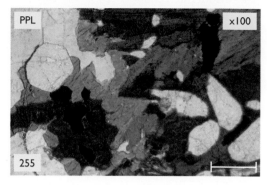

Apatite in Larvikite; Dalheim, Larvik, Norway.

Dull or no different from other c'less minerals.    19

19    Isotropic (cubic system), but some crystals are weakly anisotropic
$\delta$ = 0.000–0.001. Trapezohedral, hexagonal or granular. May show lamellar
twinning. Mafic rock: vesicle-fill or interstitial.    **ANALCITE**

Analcite in teschenite; Salisbury Crags, Edinburgh, Scotland.

Anisotropic ($\delta$ generally >0.002). Hexagonal crystals or prisms/rods. Uniaxial
interference figure. Granite/granite pegmatite.
Uniaxial –ve ($\delta$ = 0.004–0.009)    **BERYL(■)**
Uniaxial +ve ($\delta$ = 0.000–0.010)    **EUDIALYTE (■)**

Eudialyte phenocryst in alkali granite; Mt. Yukspor, Kola, Russia.

20   $1^{st}$ order grey/white to yellow/orange ($\delta = 0.004–0.015$) interf. colours.        21
     $2^{nd}$ or $3^{rd}$ ord. interf. colours; $\delta > 0.020$.                                24

21   Prismatic or columnar crystals in granites or granite pegmatites (usually isolated
     individual crystals).                                                                       22
     Fibrous/acicular or very elongate blades (aggregates); mafic rocks, vesicle fill, or
     high-T calc-silicate skarns of contact metamorphism.                                       23

22   Max. $1^{st}$ order white.                                    **EUDIALYTE(■), or BERYL(■)**

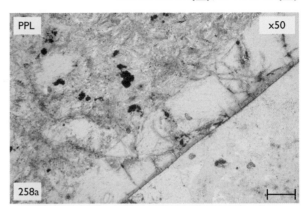

(258a) Eudialyte phenocryst in alkali granite; Mt. Yukspor, Kola, Russia;

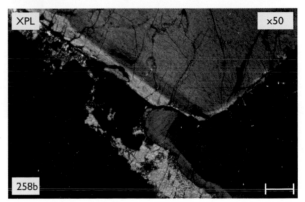

(258b) Eudialyte in sodalite-bearing syenite; Greenland.

Max. $1^{st}$ order yellow.                                                          **TOPAZ(■)**

23   Fibrous/acicular (mafic rock, vesicle-fill or interstitial). Grey to 1st ord. yell. max. interf. colours.                      **ZEOLITE (eg Natrolite, Laumontite)**

Zeolite amygdale in Tertiary dolerite dyke; Croagh Patrick, Mayo, Ireland.

Acicular/elongate rods/blades.(Calc-silicate rocks, high T metamorphism) 1st ord. yell./orange $\delta = 0.014–0.015$).                      **WOLLASTONITE**

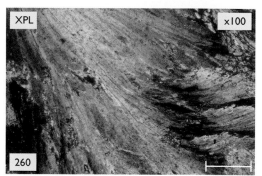

Wollastonite in calc-silicate rock in the contact aureole of the Dartmoor Granite; Meldon, Devon, England. Note the N-S oriented crystals (left) and E-W oriented crystals (centre right) showing straight extinction.

24   Low 2nd ord. (bright blue to blue-grn.) interf. colours; $\delta = 0.021–0.025$
                      **ILLITE, SMECTITE (Clay minerals) or GIBBSITE**

Andalusite porphyroblast pseudomorphed to an aggregate of clay minerals and sericite in andalusite slate; contact aureole of Skiddaw Granite, Lake District, England.

*Note: True birefringence for these minerals is usually higher (illite $\delta = 0.030$; smectite $\delta = 0.010–0.040$), but thin crystals (< 30 μm) in fine grained aggregates will give lower interference colours. X-ray diffraction will be needed to confidently identify the clay minerals (and other minerals) present in such fine grained assemblages. Gibbsite is a common mineral in laterites and bauxite.*

Mid 2nd ord. (yellow) to low 3rd ord. (blue) interf. colours; $\delta = 0.026–0.040$.   25

25   Laths (or may seem nearly acicular if fine)   (6e)                   **MUSCOVITE**

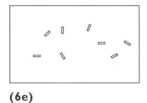

**(6e)**

Muscovite (fine-grained indistinct cleavage) in pyrite slate; locality unknown.

*Note: In fine-grained aggregates the optical properties needed to enable distinction between muscovite, paragonite and pyrophyllite will probably not be present. Muscovite is the commonest of these three minerals. It is advisable to use X-ray diffraction techniques, scanning-electron microscopy or an electron microprobe to confirm the identity of minerals in fine-grained aggregates.*

Minutely crystalline fine-grained aggregate (alteration product)   (6f).

**SERICITE (or Illite)**

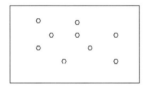

**(6f)**

Sericite alteration arrowed of plagioclase giving a fine "dusting" across the feldspar in augen gneiss; Bettyhill, Tongue, Scotland.

## LOW RELIEF

26   Isotropic:                                           **VOLCANIC GLASS (or hole in slide!)**

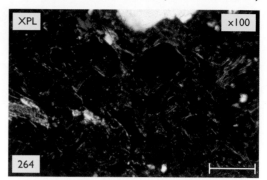

**Volcanic glass** shards in Miocene ignimbrite; Bonifacio, Corsica, France.

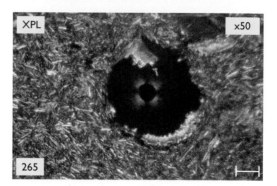

**Vesicle** in boninite; Japan.

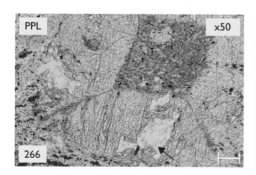

**Hole in slide** (arrowed) where a small fragment of andalusite has been plucked out during the thin-section making process. Andalusite (end-section) in crd-and slate; Connemara, Ireland.

Non-isotropic                                                                        27

27   1st order dk.grey/grey/white interf. colours                                   28
     1st order yellow/orange interf. colours                                        38

28   Fibrous/acicular crystals (6a, 4c), or thin blades; often radiating.           29
     Not fibrous/acicular or thin blades.                                           30

29   Vesicle fill or cavity fill (6g):                              **ZEOLITE (eg Thomsonite)**

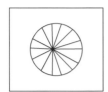

**(6g)**

Zeolite cavity-fill alteration after plagioclase in dolerite; Scawt, Antrim,
N.Ireland.

**or CHALCEDONY**

Chalcedony in basaltic andesite from
lower pillow lavas; Troodos Ophiolite,
Cyprus.

Vein mineral (6h) or in strain shadow (6i) (low T metamorphic conditions):

**fibrous QUARTZ**

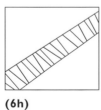

**(6h)**          **(6i)**

Fibrous quartz in strain shadow of pyrite
crystals in pyrite slate; locality unknown.

30    Phenocryst, porphyroblast, porphyroclast or clast.                31
      Groundmass, matrix, inclusions, vein-fill, alteration product.    35

31    Rectangular (qtz-absent undersaturated Na-rich ig.rock).    **NEPHELINE**

Nepheline in nephelinite; Bohemia, Czech Republic.

Not rectangular                                                          32

32    Sector-trilling/zoning or cross-hatched twinning.                 33
      No sector trilling/zoning or cross-hatched twins.                 34

33    Sector-trilling/sector zoning (**6j**). Max. interf. colours to 1$^{st}$ Order white or pale yellow ($\delta = 0.007$–$0.011$).    **CORDIERITE**

**(6j)**

Cordierite showing sector twinning in crd-and hornfels, Connemara, Ireland (see start of Section 6 for an enlargement of this image).

Complex cross-hatched twinning (**6c**). 1$^{st}$ Ord. dk.grey interf. colours ($\delta = 0.001$).    **LEUCITE**

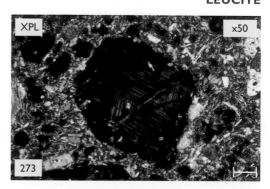

Leucite in leucite tephrite; Vesuvius, Italy.

34   Biaxial, high 2V (abundant inclusions, altered).   **PLAGIOCLASE(...)/CORDIERITE**

**Plagioclase (albite)**

Plagioclase with abundant clinozoisite inclusions in ab-czo-act mafic hornfels; locality unknown.

**Cordierite**

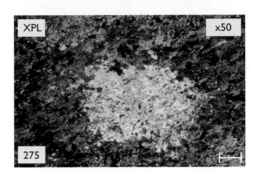

Cordierite in cordierite-andalusite hornfels, Connemara, Ireland.

Uniaxial +ve (no/few inclusions, no retrogression).   **QUARTZ**

Quartz phenocryst in recrystallised felsic dyke/sill in the vicinity of Corvock Granite, S.Mayo, Ireland.

*Note: Although plagioclase is a cleaved mineral, with biaxial interference figure, these characteristics are not always obvious in regional metamorphic rocks. Plagioclase formed in rocks with significant shear stress generally lack twinning or just show a simple twin. This makes it difficult to distinguish from cordierite and quartz. Abundant inclusions in both plagioclase and cordierite of many metamorphic rocks generally makes it impossible to obtain an interference figure. It is worth bearing in mind that quartz does not occur as porphyroblasts and that whilst both plagioclase and cordierite may show alteration to sericite, only plagioclase is likely to have abundant epidote/clinozoisite inclusions.*

35   Inclusion:                                                          **QUARTZ**

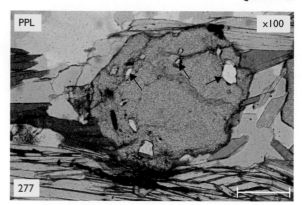

Low relief quartz inclusions (arrowed) in high relief garnet porphyroblast in garnet mica-schist; Ross of Mull, Scotland.

Not an inclusion:                                                        36

36   Felted mass of fine platy crystals (**6e, 6f**).        **KAOLINITE (clay)**

Clay minerals (and sericite) forming a pseudomorph after andalusite porphyroblasts in chiastolite slate; Skiddaw Granite aureole, Lake District, England.

Granular, polygonal, or prismatic (rectangular or rhombic) crystals.        37

37   Granular **(6b)** or polygonal aggregate **(6k)** of largely equidimensional crystals.
**QUARTZ**

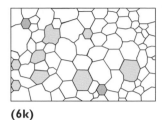

**(6k)**

(a) Polygonal aggregate of quartz in semi-pelitic schist; Mangkuma-Ketempe, Ghana.

(b) Detrital quartz grains surrounded by a calcite cement in sandstone. Hook Quarry, West Hoathly, Sussex, England.

Prismatic (rectangular to rhombic/diamond-shaped) elongate crystals.  **GYPSUM**

Gypsum in evaporite; Kirkby Thore, Cumbria, England.

38    Fibrous veinlets (ultramafic rock).                    **CHRYSOTILE**

Serpentine (chrysotile) micro–
veinlet in serpentinite; Ballantrae,
Scotland.

Vesicle-fill (mafic rocks, especially basalts); $\delta = 0.010–0.012$.

**ZEOLITE (eg Laumontite)**

Zeolite (e.g. laumontite) in meta-
dolerite/diorite; locality unknown.

# 0 Cleavage traces, coloured

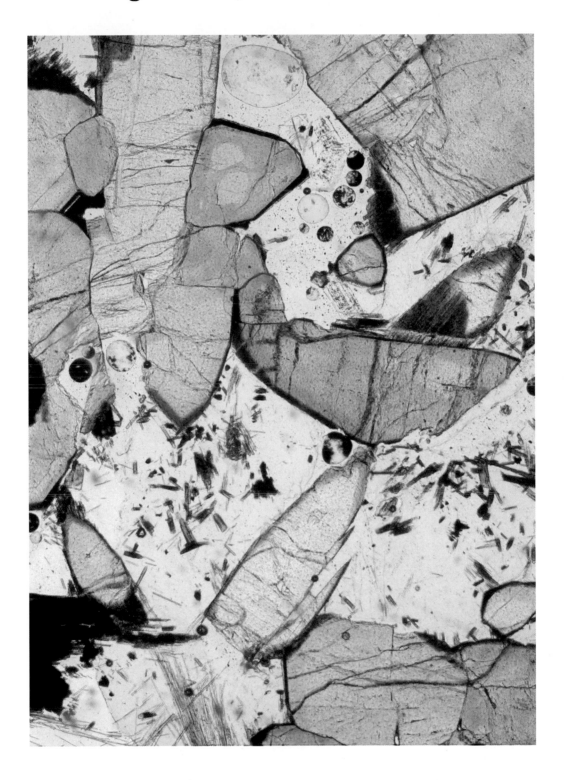

1   Extreme red to violet to yellow pleochroism.            **PIEMONTITE**

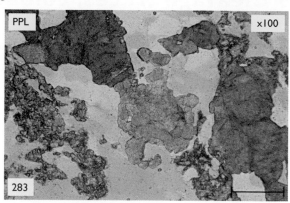

Piemontite. Pennsylvania, USA.

Not intense red to violet to yellow pleochroism.                    2

2   Blue, or colourless to pale blue/blue pleochroism                  3
    Not blue.                                                           6

**BLUE**

3   Isotropic. Medium –ve relief; anhedral or granular   (7a).

                              **SODALITE Gp. (e.g Sodalite, Lazurite)**

**(7a)**

Lazurite (sodalite group) in metasomatised marble; locality unknown.

Anisotropic.                                                          4

4  High relief                                                    5
   Moderate relief. ($\delta = 0.009-0.022$); small needles    (**7b**).

**GLAUCOPHANE(...)**

**(7b)**

Glaucophane (needles) as inclusions in garnet within glaucophane-garnet schist (blueschist); Ile de Groix, Brittany, France.

5  1$^{st}$  order  grey/white  interf.  colours  ($\delta$ = 0.005-0.008).  Granulite  facies metamorphism.

**SAPPHIRINE**

Sapphirine in sapphirine-biotite rock; locality unknown.

Top 1$^{st}$ ord./ low 2$^{nd}$ ord. interf. colours ($\delta = 0.011-0.020$).    **DUMORTIERITE**

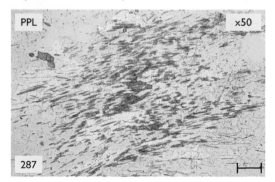

Dumortierite in dumortierite-qtz rock; locality unknown (also see Image 198, and an enlargement of this image at the start of Section 4).

6  Yellow, orangy-yell., or c'less to yellow or yell.-grn. pleochroism.          7
   Green, brown, pink/red, red-brn. or orange-brn.                               11

   **YELLOW (or yellow-green, or orange)**

7  Low to medium relief; fine grained aggregate.
                                    **PALAGONITE/ SMECTITE (a clay mineral)**

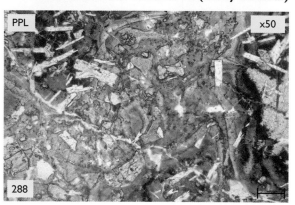

Palagonite in palagonite tuff; Portree, Skye, Scotland.

   Medium to high relief                                                          8

8  Max. 1$^{st}$ ord. yell./orange interf. colours ($\delta = 0.011$-$0.014$), high relief (metapelite); colourless to yellow pleochroism.                                 **STAUROLITE**

Staurolite in stt-bt-schist; Glen Esk, Scotland (see Images 193 and 212 for staurolite in PPL).

   2$^{nd}$/3$^{rd}$ order interference colours                                    9

9   Max. low/mid 2$^{nd}$ ord. interf. colours ($\delta = 0.021\text{-}0.029$).

**TOURMALINE (Dravite)**

Tourmaline (Dravite); locality unknown.

Max. 3$^{rd}$ ord. interf. colours.                                                    10

10   Yellow to yellow-green, colourless to yellow-green pleochroism, ($\delta = 0.015\text{--}0.051$). High relief. Biaxial –ve, high 2V = 90–116°. Common mineral, especially in metabasites and alteration zones.

**EPIDOTE**

Note:  *Strength of colour in epidote increases with increased $Fe^{3+}$ in epidote (= pistacite content).*

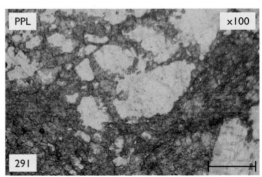

Epidote in epidotised granite; Ghana.

Yell.-orangy yell. or c'less/pale yellow pleochroism, ($\delta = 0.026\text{-}0.041$). Moderate to high relief. Biaxial +ve, 2V c.50-80°. Rather uncommon mineral of certain calc-silicate rocks and skarns in mineralised areas.

**HUMITE Gp. (eg Humite, Clinohumite, Chondrodite)**

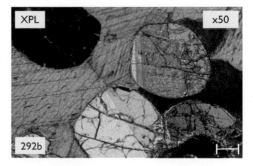

Chondrodite in mineralised skarn; Ottawa, Canada. Note: Clinohumite and chondrodite may show lamellar twinning.

11  Green to blue-grn.                                                       12
    Brown, red-brn., pink/red, orange-brn. or yell.-brn.                     18

**GREEN or BLUE-GREEN**

12  High relief                                                              13
    Low or moderate relief                                                   15

13  Isotropic:                                                          **SPINEL**

*Note:* **SPINEL** sensu stricto *is the Mg end-member of the Spinel series, and* HERCYNITE *the Fe end-member. However,* Spinel sensu lato *shows a signif-icant amount of Fe$^{2+}$ substitution for Mg. This mineral is called* **PLEONASTE** *(dark green), and is associated with high temperature metamorphism.*

Spinel (pleonaste) in crn-spl-mag rock; local-ity unknown.

Non-isotropic                                                                14

14  Mid/upper 2$^{nd}$ order interf. colours, strong pleochroism (± colour zonation), // extinction. Yell.-grn-green-blue-grn. pleochroism.     **TOURMALINE (Schorl)**

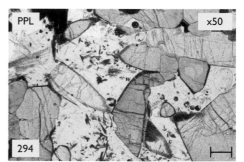

Tourmaline (schorl) in quartz-tourmaline rock near granite intrusion; Roche, Cornwall, England (see start of Section 7 for an enlarge-ment of this image".

Upper 1$^{st}$ order (or anomalous) interf. colours, weak or moderate c'less-yell.grn.-yell.-brn.-blue-grn. pleochroism.                            **PUMPELLYITE**

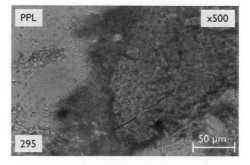

Pumpellyite (highly magnified) in sub-greenschist facies metavolcanic rock; Rose-land, Cornwall, England.

15    Moderate relief                                                          16
      Low relief                                                               17

16    Colourless to green or pale/dark green pleochroism, 1st order grey (or anomalous)
      interf. colours:                                                    **CHLORITE**

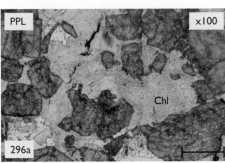

Chlorite (with and without obvious cleavage) in cassiterite-rich vein rock within granite;
East Pool and Agar Mine, Cornwall, England.

*Note:* The subhedral high relief brown crystals (PPL) are the tin ore mineral, cassiterite.

Strongly coloured, sometimes yell-green to deep green pleochroism; top 1st order/
mid 2nd order interf.colours, but generally masked by the strong colour of the
mineral (sandstone/greensand/impure limestone).                    **GLAUCONITE**

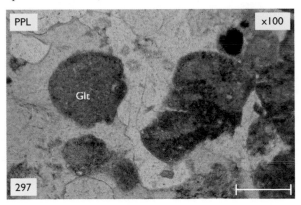

Glauconite in Cretaceous green-
sand; Folkestone, Kent, England.

17    Alteration of olivine (mafic/ultramafic rock or marble).

**SERPENTINE** *(e.g. Antigorite)*

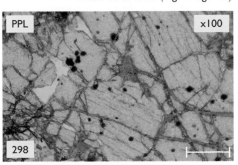

Serpentine micro-veinlets formed as part of
olivine alteration in picrodolerite; Duntulm,
Skye, Scotland.

Alteration of cordierite (metasediment/high T met. rock).                    **'PINITE'**

*Note: Pinite is a microcrystalline aggregate principally of muscovite (sericite) and chlorite, which gives the green colour.*

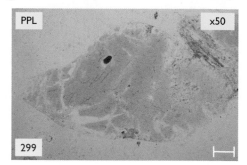

Pinnitised cordierite in cordierite-bearing quartzite; Kragerö, Norway.

## BROWN/ RED/ PINK/ORANGE-BRN./ YELL.-BRN.

| | | |
|---|---|---|
| 18 | High/very high relief | 19 |
| | Low or Moderate relief | 29 |
| 19 | Isotropic (or appearing to be) | 20 |
| | Anisotropic | 24 |
| 20 | Well defined grains, often with good crystal form | 21 |
| | Fine grained or amorphous mass | **LIMONITE** |

*Note: Limonite (yellow, yell.-brn. to orange in PPL) is a crypto-crystalline mix of Fe-oxides/hydroxides (goethite, lepidocrocite, ± haematite).*

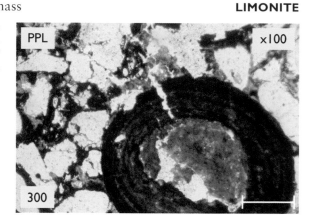

Limonite in oolitic limestone; Abbotsbury, Dorset, England.

21  Needle-like or rodded crystals                                    22
    Anhedral, rounded, hexagonal, square or other form            23

22  Appears almost black in PPL, due to strong absorption, but thin crystal edges are brown to dk. brown (blueschist facies meta-ironstone).    **DEERITE**

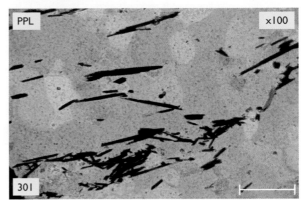

Deerite in stilpnomelane-rich siliceous meta-ironstone; Laytonville Quarry, CA, USA.

Essentially opaque in thin-section, but very thin edges of crystals may show bright red in PPL.    **HAEMATITE (Specular Haematite)**

*Note: Specular haematite has a "scaly" or "micaceous" crystal form. Cross-sections of such thin platy crystals exhibit needle-like or square-ended rod form in thin-section*

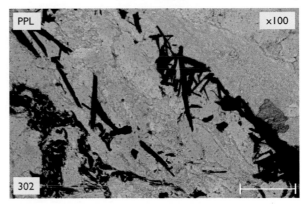

Specular haematite in haematite-quartz-sericite vein; Cap de la Hague, Normandy, France.

23   Brown to dark brown (igneous rocks)                **PEROVSKITE**

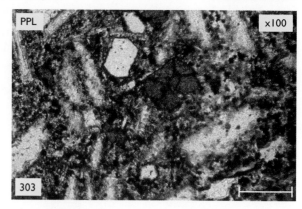

*Note: Ti-garnets (**Melanite** and **Schorlomite**) of certain alkaline igneous rocks are also brown in thin-section. Also check out AENIGMATITE, which is v.dark brown, pleochroic, but can appear isotropic.*

Perovskite in olivine melilitite; Katunga, Burundi.

Pale pinkish/pale brownish                **ALMANDINE (garnet)**

Garnet (almandine-rich; c. $Alm_{70\%}$) in regional metamorphic garnet mica-schist; Garve, Scotland.

Pale brown                          **SPESSARTINE (garnet)**

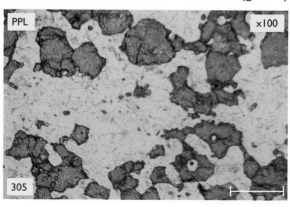

Spessartine garnet in hydrothermally altered quartzo-feldspathic metavolcanic rock; Nisserdal region, Telemark, Norway (also see Images 87–88).

24   Very weak birefringence ($\delta$ = 0.000-0.002) 1$^{st}$ Ord. dark grey to black. Larger crystals may show complex (cross-hatched) twinning.   **PEROVSKITE**

*Note: Also check out AENIGMATITE, which is v.dark brown, pleochroic, and despite high birefringence ($\delta$ = 0.070-0.080), this is masked by strong background colour so impossible to determine.*

Perovskite (showing very low first-order interference colours and complex twinning) with olivine in peridotite; Africanda, Kola Peninsula, Russia. (photo courtesy of Giles Droop).

Birefringence > 0.003 (1$^{st}$ Ord mid grey or higher). Twinning rare and, if present, is simple.   25

25   Needle-like (**7b**) or rodded crystals (and granular) with strong red-brown colour. V.high relief.   **RUTILE**

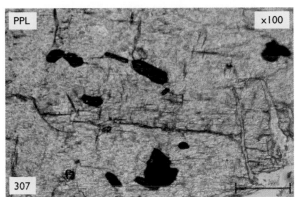

Rutile (small rods) included in kyanite, within kyanite schist; Loch Assapol, Mull, Scotland.

Granular rounded/oval (**7a**), or hexagonal (**7c**), or polygonal, diamond-shaped or amorphous, or sphaerulitic aggregates.   26

(**7c**)

26   Relief varies on rotation (**5a**) (moderate to high to v.high). May occur as sphaeru-
     litic aggregates:                                                    **SIDERITE**

Siderite (sphaerulitic form) in Jurassic
sideritic sandstone; Yorkshire, England.

Relief stays constant on rotation                                              27

27   Pale brn.- yell.brn., 4th/5th ord. pastel interf. colours:     **TITANITE (=Sphene)**

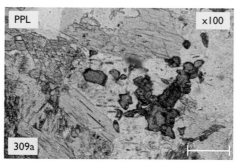 

Titanite in actinolite skarn; Loch Leuh, Donegal, Ireland.

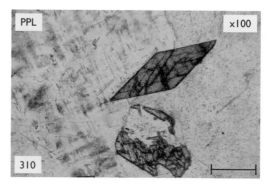

Diamond-shaped crystal of titanite
(=sphene) with adjacent crystal of spodumene set in a background of microcline.
Spodumene granite; Leinster, Ireland.

Deep brown, red-brown or red.                                                  28

28   Individual granular, round/oval crystals (**7b**). v.high relief (RI = 2.61-2.90).

**RUTILE**

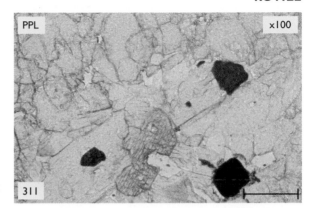

Rutile in kyanite eclogite; Qinling Belt, Dabie Shan, China.

Amorphous mass (alteration of olivine, especially in basalt):    **IDDINGSITE**

*Note:  Iddingsite is a fine grained*
*mix of smectite, chlorite and*
*goethite/haematite.*

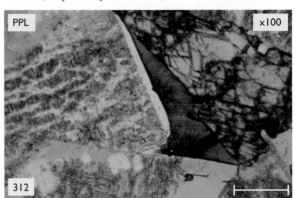

Iddingsite in ekerite; Drammen, Norway.

29   Moderate relief, roughly hexagonal (7c) or subhedral. Dk.brown or red-brn (non-pleochroic), birefr. nil, low or masked.                                    **BIOTITE(■)**

(a) Biotite end-section (hexagonal no pleochroism, no cleavage) and more typical side-section (sub-rectangular pleochroic lath with cleavage) in granite; St.Jacut, Brittany, France. The biotite crystals also display dark pleochroic haloes around minute zircon inclusions (see Image 251).

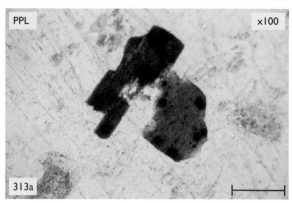

(b) Biotite end-section (in granite) showing miniscule needles of rutile with a crystallographically-controlled arrangement to give a pattern at 60° angles in a fine 'Widmanstatten-type' exsolution texture (see Spry, 1969, pg.98). This does not represent intersecting cleavages.; St. Jacut, Brittany, France.

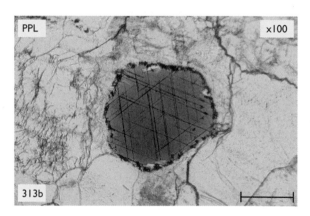

Low or mod. relief. Amorphous yellow-orange/yell.-brn. alteration product.
                                                              **PALAGONITE/SMECTITE**

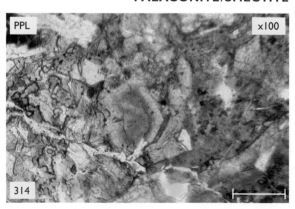

Palagonite in palagonite tuff; Portree, Skye, Scotland (also see Image 288).

# Opaque minerals

## OPAQUE MINERALS

As the present key is designed for examination of standard 30μm thin-sections with a glass cover slip, using a transmitted light petrological microscope, it is not suited to the detailed identification of opaque minerals. For accurate identification of such phases, it is advisable to use thicker, polished thin-section (without cover slip) and to study using a metallurgical reflected light microscope. For further detail on the identification of such phases, refer to an ore mineralogy text (e.g. Craig & Vaughan, 1994, or Ixer, 1990).

Despite the limitations of opaque mineral identification when not using a reflected light microscope, it is still possible to make some reasonable mineral interpretations in a standard thin-section based on shape, form and associated minerals. This final Section provides a key to some of the commonest opaque minerals, seen as accessory phases in igneous, metamorphic and sedimentary rocks, and also as common vein or matrix phases in rocks of mineralised districts.

As with previous Sections of this key, the emphasis is on features seen during standard transmitted light microscopy, but colours in reflected light (RL) are also mentioned for those who wish to undertake more thorough examination. If the microscope you are using has a reflected light option, you could divert or turn off the transmitted light and switch to reflected light mode, to see the basic RL colour for the mineral of interest. If the crystals of interest are of reasonable grain size, studying the thin-section in good light with a hand lens will prove very useful. Alternatively, a standard binocular microscope used for hand specimen work or biological samples is very effective. The observations are best made with an oblique reflected light source.

All images shown below were taken either using transmitted light (PPL or XPL) with a standard petrological microscope, or else with the transmitted light source turned off and an oblique reflected light source simply positioned to one side of the microscope stage and shone from above at about 45° to the specimen.

The full range of opaque minerals associated with mineralised areas is outside the scope of this key, but it is hoped that the approach suggested above, and the images presented below, will prove helpful in enabling some of the common opaque minerals that occur as accessory phases in many rocks to be recognised.

## OPAQUE MINERALS

1  Euhedral, subhedral, anhedral or skeletal         2
    Fine grained, cryptocrystalline or amorphous      8

2  Acicular/rodded, or elongate laths (cross-sections through platy/scaly crystals) (**8a**).      3
    Squares, rhombs, diamonds, pentagons, hexagons, triangles, and/or octahedra (**8b**), or anhedral/rounded (**8c**) or skeletal (**8d**).      4

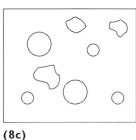

**(8a)**

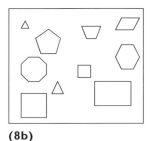

**(8b)**

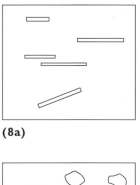

**(8c)**

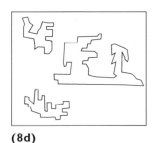

**(8d)**

3  Elongate laths (**8a**). Thin edges may be brown in PPL. V.high relief (RI = c.2.70). RL = Brown with pink/violet tint (darker than haematite). Very common in regional metamorphic rocks.    **ILMENITE**

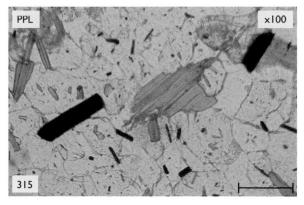

Ilmentite in granoblastic crd-bt-ilm rock; locality unknown.

Elongate laths (**8a**) and needles. Thin margins may be dark red in PPL. V.high relief (RI = 2.87–3.22). RL = Grey-white, (with blueish tint).

**SPECULAR HAEMATITE**

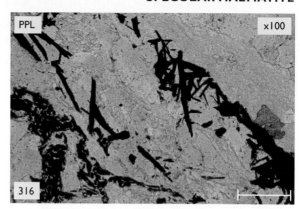

Specular haematite in haematite-quartz-sericite vein; Cap de la Hague, Normandy, France.

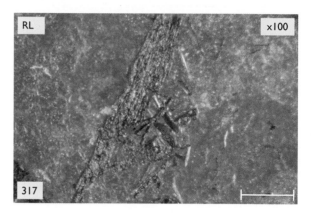

Specular haematite in haematite-quartz-sericite vein; Cap de la Hague, Normandy, France.

4   Crystals exhibit elongate triangular pits. Occurs as anhedral crystals or euhedral cubes. RL = white (but usually mid grey when viewed with cover slip on thin-section in oblique RL). Associated with Pb-mineralised areas.    **GALENA**

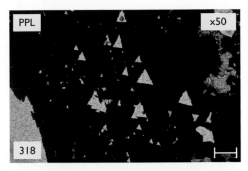

Galena in a Zn-mineralised siliceous rock; Minera, Wrexham, Wales. The triangular pits form where material has been plucked out in thin-section preparation. These elongate triangles are picking out cleavage of this cubic mineral.

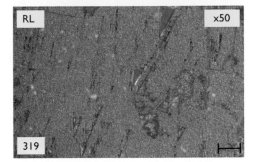

Crystals do not exhibit elongate triangular pits                                                5

5   Euhedral or subhedral crystals (8b), or round (8c)                                          6
    Skeletal crystals (8d)                                      **ILMENITE (or Ilmenomagnetite)**

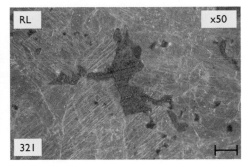

Ilmenite (or ilmenomagnetite) in gabbro; Beinn Bheag, Mull, Scotland. Skeletal ilmenite crystals are common in mafic igneous rocks. Thin edges may be brown in PPL. V.high relief (RI = c.2.70).

6   Squares, rhombs, hexagons, triangles. RL = pale yellow (will look brassy or golden when thin-section viewed with hand lens). **PYRITE**

Pyrite in slate; locality unknown (see start of Section 8 for an enlargement of this image).

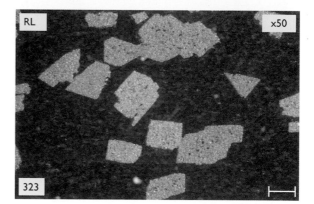

Pyrite in slate; locality unknown. Pyrite looks brassy or golden when thin-section is held up to light and viewed through a hand lens. In RL (polished sections), pyrite is pale yellow.

Round, octagonal, diamond-shaped, square, or hexagonal. RL = dk.grey to black (will look black or steely metallic dark grey with hand lens).    7

7   Magnetic. If abundant in the thin-section it has a strong magnetic effect. To test this, place a compass on a flat surface and glide the thin-section slowly over it. If magnetic, the compass needle will be deflected. Magnetite appears steely metallic dk. grey when observed through a hand lens in oblique reflected light. RL = dk. grey with brown tint. Common accessory mineral in many igneous and metamorphic rocks and a detrital mineral in sedimentary rocks.   **MAGNETITE**

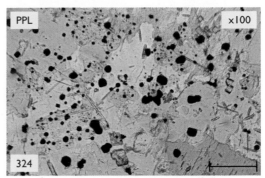

Magnetite in sil-bt-kfs-mag hornfels; Weets, Grampian, Scotland.

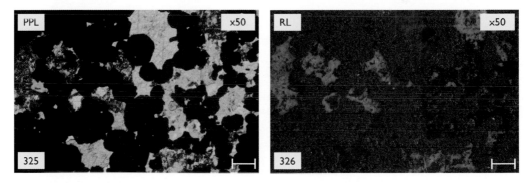

(325-326) Magnetite in monticellite-magnetite skarn; Camas Malag, Skye, Scotland. Being a magnetic mineral, high concentrations, such as this example, are strongly magnetic.

Non-magnetic. High concentrations of chromite will not deflect a compass needle in the manner described above for magnetite. RL = Dark grey to brownish grey. Associated with ultramafic igneous rocks (e.g. peridotites), metamorphosed ultramafic rocks (e.g. serpentinites). Appears dk.grey to black through the hand lens.                                                                                                **CHROMITE**

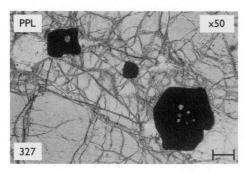 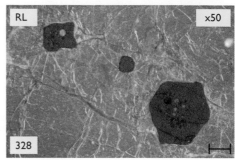

Chromite in olivine-rich ultrabasic rock of cumulate sequence; Rhum, Scotland.

Chromite in 1cm thick chromitite layer with anorthitic plagioclase making up the background; within ultramafic cumulate sequence, Rhum, Scotland.

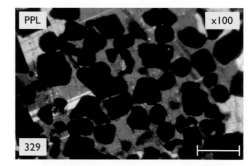

8   Fine grained cryptocrystalline or amorphous opaques cannot be identified with confidence in thin-section by standard transmitted light microscopy. To correctly identify these minerals, reflected light microscopy and electron microprobe analysis or scanning-electron microscopy will be needed. In many mudrocks and carbonates/calc-pelites, finely disseminated carbonaceous material is usually present. When metamorphosed (e.g. schists, marbles), the fine carbon develops a more defined mineral structure as it recrystallises to graphite. However, even this is only aggregates of fine specks. RL = Brn.-gry to dk.grey-black bireflectance pleochroism.                                                                              **GRAPHITE**

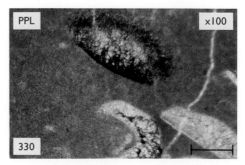

Carbonaceous material in biomicrite limestone; Purbeck, Dorset, England.

# Appendix 1: Birefringence value (δ) and corresponding interference colours for a 30 μm rock thin-section. The "phase difference" or "retardation value" (in nm) marking the top of each order is given in the first column.

|  |  |  | δ |
|---|---|---|---|
| Isotropic |  | black | 0.000 |
| 1st Order | Low | grey (or 'Berlin blue') | 0.003–0.005 |
|  |  | pale grey/white | 0.006–0.008 |
|  | Mid | pale yellow/yellow | 0.009–0.012 |
|  |  | orange | 0.013–0.015 |
| 550 nm | Top | red/deep red | 0.016–0.018 |
| 2nd Order | Low | violet/blue | 0.019–0.021 |
|  |  | blue-green | 0.021-0.023 |
|  |  | green/yell.grn | 0.023–0.025 |
|  | Mid | yellow | 0.025–0.027 |
|  |  | orange | 0.027–0.031 |
|  |  | red | 0.031–0.034 |
| 1101nm | Top | purple/violet | 0.034–0.037 |
| 3rd Order | Low | violet/blue/blue-grn | 0.037–0.038 |
|  |  | green | 0.038–0.043 |
|  |  | yellow-peach | 0.043–0.047 |
| 1652 nm | High | pink/pale pink | 0.047–0.055 |
| 4th Order | Low | pale green | 0.055–0.064 |
|  |  | pale cream/peach | 0.064–0.068 |
| 2203 nm | High | v.pale pink | 0.068–0.075 |
| 5th Order+ |  | v.pale pastel shades, cream, pearl grey to milky white | 0.075–0.200 |

# Appendix 2: Birefringence table of minerals in order of δ values (based on data in Deer *et al.*, 1992, 2013, and Kerr, 1977)

| | |
|---|---|
| Fluorite | Isotropic |
| Spinel (e.g. Pleonaste) | Isotropic |
| Garnet (Almandine) | Isotropic |
| Garnet (Pyrope) | Isotropic |
| Volcanic glass, and glass (hole in slide) | Isotropic |
| Lazurite | Isotropic |
| Limonite | Isotropic |
| Periclase | Isotropic |
| Sodalite Group (Sodalite, Nosean, Haüyne) | Isotropic |
| Sphalerite | Isotropic |
| Spinel | Isotropic |
| Analcite | Isotropic (to 0.001 in some cases) |
| Garnet (Spessartine) | Isotropic (to 0.001 in some cases) |
| Garnet (Andradite) | Isotropic (to 0.005 in some cases) |
| Collophane | Isotropic (to 0.005 in some cases) |
| Garnet (Grossular) | Isotropic (to 0.008 in some cases) |
| | |
| Palagonite | 0.000–0.001 |
| Perovskite | 0.000–0.002 |
| Zeolite | 0.000–0.015 |
| Mesolite | 0.001 |
| Leucite | 0.001 |
| Apatite | 0.001–0.007 |
| Vesuvianite (Idocrase) | 0.001–0.009 |
| Chlorite | 0.001–0.010 |
| Melilite | 0.001–0.013 |
| Chabazite | 0.002–0.010 |
| Nepheline | 0.003–0.005 |
| Zoisite | 0.003–0.008 |
| Eudialyte | 0.003–0.022 |
| Serpentine (Antigorite) | 0.004–0.007 |
| Beryl | 0.004–0.009 |
| Clinozoisite | 0.004–0.015 |
| Scapolite (Na-rich [Marialite], $Me_0–Me_{50}$) | 0.004–0.020 |
| Sapphirine | 0.005–0.009 |

| | |
|---|---|
| Chloritoid | 0.005–0.022 |
| Clay (Kaolinite Gp.) | 0.006 |
| Stilbite | 0.006–0.008 |
| Orthoclase | 0.006–0.010 |
| Microcline | 0.006–0.010 |
| Antiperthite | 0.006–0.010 |
| Anorthoclase | 0.006–0.010 |
| Sanidine | 0.006–0.010 |
| Perthite (Microperthite) | 0.006–0.010 |
| Thomsonite | 0.006–0.012 |
| Riebeckite | 0.006–0.016 |
| Jadeite | 0.006–0.021 |
| Heulandite | 0.007 |
| Scolectite | 0.007 |
| Plagioclase (Albite-Labradorite) | 0.007–0.010 |
| Adularia | 0.008 |
| Corundum | 0.008–0.009 |
| Topaz | 0.008–0.011 |
| Enstatite | 0.008–0.014 |
| Cordierite | 0.008–0.018 |
| Crossite | 0.008–0.020 |
| Quartz | 0.009 |
| Celestine | 0.009 |
| Chalcedony | 0.009 |
| Andalusite | 0.009–0.012 |
| Plagioclase (Bytownite) | 0.009–0.012 |
| Axinite | 0.009–0.016 |
| Gypsum | 0.010 |
| Arfvedsonite | 0.010–0.012 |
| Hypersthene | 0.010–0.016 |
| Pumpellyite | 0.010–0.020 |
| Clay (Smectite Gp.) | 0.010–0.040 |
| Staurolite | 0.011–0.014 |
| Dumortierite | 0.011–0.020 |
| Barytes | 0.012 |
| Laumontite | 0.012 |
| Plagioclase (Anorthite) | 0.012–0.013 |
| Natrolite | 0.012–0.013 |
| Kyanite | 0.012–0.016 |
| Monticellite | 0.012–0.020 |
| Brucite | 0.012–0.020 |
| Tremolite | 0.012–0.027 |
| Omphacite | 0.012–0.028 |
| Wollastonite | 0.013–0.014 |
| Serpentine (Chrysotile) | 0.013–0.017 |
| Glaucophane | 0.013–0.020 |
| Anthophyllite | 0.013–0.026 |

| | |
|---|---|
| Allanite (*Orthite*) | 0.013–0.036 |
| Actinolite | 0.014–0.026 |
| Spodumene | 0.014–0.027 |
| Ferrosilite | 0.015–0.022 |
| Epidote | 0.015–0.051 |
| Hornblende | 0.016–0.028 |
| Cummingtonite | 0.016–0.045 |
| Grunerite | 0.016–0.045 |
| Sillimanite | 0.018–0.022 |
| Augite | 0.018–0.033 |
| Titanaugite | 0.018–0.033 |
| Lepidolite | 0.018–0.038 |
| Lawsonite | 0.019–0.021 |
| Glauconite | 0.020–0.030 |
| Scapolite (Ca-rich [Meionite], $Me_{50}$–$Me_{100}$) | 0.020–0.038 |
| Tourmaline (Dravite) | 0.021–0.029 |
| Prehnite | 0.022–0.035 |
| Hemimorphite | 0.022–0.035 |
| Pigeonite | 0.023–0.029 |
| Diopside | 0.024–0.031 |
| Cancrinite | 0.025 |
| Hedenbergite | 0.025–0.034 |
| Tourmaline (Schorl) | 0.025–0.035 |
| Piemontite | 0.025–0.073 |
| Yoderite | 0.026 |
| Chondrodite | 0.028–0.034 |
| Paragonite | 0.028–0.038 |
| Clinohumite | 0.028–0.041 |
| Kaersutite | 0.028–0.047 |
| Humite | 0.029–0.031 |
| Deerite | 0.030 |
| Clay (Illite) | 0.030–0.035 |
| Aegirine-Augite | 0.030–0.050 |
| Stilpnomelane | 0.030–0.110 |
| Biotite | 0.033–0.059 |
| Olivine (Forsterite) | 0.035 |
| Muscovite | 0.036–0.049 |
| Iddingsite | 0.038–0.044 |
| Anhydrite | 0.040 |
| Diaspore | 0.040–0.050 |
| Aegirine | 0.040–0.060 |
| Zircon | 0.042–0.065 |
| Phlogopite | 0.044–0.047 |
| Monazite | 0.045–0.075 |
| Talc | 0.050 |
| Pyrophyllite | 0.050 |
| Olivine (Fayalite) | 0.052 |

| | |
|---|---|
| Astrophyllite | 0.060 |
| Titanite (Sphene) | 0.100–0.192 |
| Goethite | 0.138–0.140 |
| Aragonite | 0.155–0.156 |
| Calcite | 0.172–(0.190) |
| Dolomite | 0.179–(0.185) |
| Ankerite | 0.182–0.202 |
| Magnesite | 0.190–(0.218) |
| Rhodochrosite | (0.190)–0.219–(0.230) |
| Siderite | (0.207)–0.242 |
| Rutile | 0.286–0.296 |
| Lepidocrocite | 0.570 |

# Appendix 3: Glossary

(based largely on Barker (1998), with additions and amendments)

**2V angle**  The acute angle between optic axes of biaxial minerals (also referred to as the optic axial angle). This angle is bisected by the $\gamma$ optic axis, which in this context is referred to as the acute bisectrix (see 1.4.6 and Fig. 1.27).

**Acicular**  A term used to describe crystals with needle-like form (see Fig. 1.3b).

**Acute bisectrix**  (see 2V angle)

**Amphibolite**  A metamorphosed basic igneous rock with a mineral assemblage comprised largely of amphibole and plagioclase, usually with quartz and epidote.

**Amygdale**  A term for vesicles infilled by a secondary mineral (e.g. zeolites, calcite) (see Image 236a).

**Anatexis**  The process of partial melting of high grade metamorphic rocks in the presence of $H_2O$. This process produces granitoid melts and typically operates in the middle and lower crust during orogeny.

**Anhedral**  Crystals with irregular shape, lacking any of their characteristic faces (see Fig. 1.3d).

**Anisotropic**  Minerals with different optical properties (e.g. refractive indices) in different directions are said to be optically anisotropic (opposite of isotropic). Anisotropic minerals show birefringence in cross-polarised light producing interference colours caused by anisotropy affecting how different light rays pass through the crystal.

**Antiperthite**  A feldspar intergrowth comprising K-feldspar inclusions enclosed within plagioclase (the converse of perthite).

**Becke line**  Best seen at high magnification (x400) with the sub-stage condenser raised and the diaphragm closed, the Becke Line is a bright line of light concentration at the boundary between minerals of different relief (relating to different refractive indices). As the microscope tube is raised (slightly defocused), the Becke line moves towards the higher relief mineral. Named after the German mineralogist F. Becke who first used the technique in 1893 (see Fig. 1.8).

**Bertrand lens**  An accessory lens on the petrological microscope, situated between the objective and the eye-piece, used in conoscopic illumination to bring the interference figure on the surface of the objective lens into focus with the eye-piece.

**Biaxial interference figure**  In conoscopic illumination, the light passing through a mineral specimen diverges on exit to produce a conical beam. In biaxial minerals (e.g. muscovite), cut perpendicular to the acute bisectrix, this beam gives an interference figure on the surface of the objective lens, comprising two curved black lines ("isogyres"). By insertion of a Bertrand lens, the interference figure is projected to the eye-piece (see 1.4.6 and Fig. 1.26).

**Birefringence (δ)**   A numerical value defined by the difference between maximum and minimum refractive indices for a particular mineral. These differences are consistent with specific crystallographic axes. The magnitude of birefringence is reflected by the interference colours produced in cross-polarised light (see 1.4.1, Fig. 1.14 and Appendices 1 & 2).

**Blueschist**   A metamorphosed mafic rock indicative of high-P/low-T subduction-related metamorphism. It contains large quantities of sodic (blue) amphibole (glaucophane/crossite), and has a pronounced schistosity (see Image 196).

**Calc-silicate rock**   A rock with a chemistry dominated by calcium and silica, consisting of hydrous or anhydrous calc-silicate minerals such as tremolite, diopside and grossular. Carbonate minerals are also commonly present (see Images 137 & 141).

**Chemical Zoning**   Regular or abrupt changes in mineral chemistry from core to rim (see Fig. 1.18b).

**Cleavage (of minerals)**   Crystallographically defined planes of preferred splitting, corresponding to planes of weaker bond-strength in the structure of the mineral. Their arrangement parallels particular crystal faces of the mineral, in turn related to the crystal system. In thin-section, cleavage traces appear as a series of parallel lines. Some minerals have no cleavage whilst others show one or more cleavages (see Fig. 1.5).

**Conoscopic illumination**   Interference figures can only be produced in conoscopic illumination (= convergent polarised light). This is achieved by raising the sub-stage condenser to just below the microscope stage, then flipping the front lens of the sub-stage condenser "in", to converge the light on a small central area of the thin-section. On leaving the thin-section the light diverges to give a slightly conical beam.

**Contact aureole**   The zone of rocks surrounding a plutonic igneous intrusion which are thermally metamorphosed in response to the intrusion. The contact aureole can usually be sub-divided into a series of concentric zones of varying metamorphic grade based on mineral assemblages.

**Contact metamorphism**   Thermal metamorphism associated with igneous intrusions, often accompanied by fluid infiltration. Such metamorphism produces a series of metamorphic reactions in the surrounding country rock to give a contact aureole.

**Corona**   A monomineralic or polymineralic rim totally surrounding a core of another mineral phase. It represents an arrested reaction between the core phase and other components.

**Cross-hatched twinning**   'Cross-hatched twinning', 'tartan pattern' or 'gridiron twinning' is one of the most characteristic features of microcline. It results from the high-angle intersection of multiple albite and pericline twins (see Fig. 1.17c).

**Crystallographic axes**   The reference axes used to define the three-dimensional geometry of minerals. The length and angular relationships between these axes vary according to their crystal system (see Fig.1.1). To avoid any confusion, please note that the present author has used the x,y,z notation for consistency with Deer et al. (1992, 2013), but in other texts (e.g. Kerr, 1977; Ehlers 1987a,b), the principal crystallographic axes are labelled a,b,c.

**Deformation bands**   Distinct bands of deformation in crystals. Optically they are more sharply defined than undulose extinction, and represent higher strain. They can terminate either at grain boundaries or inside grains.

**Deformation lamellae**   Sets of very narrow planar or tapered lenticular structures in quartz and many other silicates. They terminate inside grain boundaries, and show a variety of orientations, though are frequently perpendicular to bands of undulose extinction.

**Deformation twins**   Twins developed in crystals in response to deformation. Especially common in calcite (Fig. 1.17d), plagioclase and pyroxene. Deformation twins terminating within the crystal have tapered ends.

**Dichotomous key**   A key used to identify something (e.g. plant, animal, mineral) based on a series of steps (couplets) of dichotomous questions (i.e questions for which there can be only one of two answers). The questions present a choice of distinguishing characters, and depending on the answer given, the user either progresses to the next question, or else is directed to another part of the key. The process continues until the identity is obtained.

**Disequilibrium**   An incompatible association of mineral phases and/or a combination of textures and structures incompatible with prevailing conditions.

**Eclogite**   A dense, high grade metamorphic rock of mafic composition with an essential assemblage of clinopyroxene (omphacite) and garnet.

**Epitaxial growth**   The oriented growth of one mineral phase in optical continuity with another due to structural similarities between the two phases. (e.g. growth of secondary amphibole rims on earlier amphiboles of different composition).

**Equilibrium**   That state of a rock system in which the phases present are in the most stable, low energy arrangement, and where all phases are compatible with the given P, T, and fluid conditions.

**Euhedral**   Crystals with good shape, completely bounded by their characteristic faces (see Fig. 1.3a).

**Exsolution**   The process whereby an initially homogeneous solid solution separates into two (or possibly more) distinct crystalline phases (typically during cooling) without the addition or removal of material (modified after Bates & Jackson, 1980).

**Extinction angle**   For minerals with straight (=parallel) extinction, there is parallelism between the crystallographic axes and optic axes (vibration directions). However, minerals showing inclined extinction (e.g. clinopyroxenes, clinoamphiboles), have varying levels of angular difference between the optic axes and the crystallographic axes (see Fig. 1.20). The extinction angle of a crystal is obtained (in XPL) by measuring the angular difference between the long-axis of the crystal when aligned with N-S cross-wires, and the position oblique to this where the crystal goes into extinction (see Fig. 1.21-1.22). By convention, the angle quoted is the smaller value with respect to N-S (ie < 45°).

**Fibrolite**   A fine fibrous or hair-like variety of sillimanite common in amphibolite facies schists and gneisses (see Image 244). Often occurs in matted aggregates or knots termed *faserkiesel*.

**Fluid inclusion**   A term for microscopic and sub-microscopic (and rarely macroscopic) inclusions of fluid trapped in minerals during primary crystallisation or fracture healing. They are typically < 50 μm in size.

**Foliation**   A set of closely-spaced planar surfaces produced in a rock as a result of deformation (e.g. schistosity, cleavage).

**Gneiss**   A coarsely banded high grade metamorphic rock consisting of alternating, mineralogically distinct (usually felsic and mafic) layers.

**Granoblastic structure**   An aggregate consisting of equidimensional crystals of approximately equal sizes. In many cases the crystals are rounded to anhedral, but granoblastic-polygonal aggregates are equally common. Granoblastic structure is especially characteristic of granulites, eclogites and many hornfelses.

**Granulite**   A high grade metamorphic rock typically with granoblastic structure and with an essential assemblage of pyroxene (typically hypersthene) and anorthitic plagioclase (see Image 190).

**Greenschist**   A low grade mafic rock with schistose texture and a mineral assemblage consisting largely of actinolite, chlorite, epidote, albite, quartz, and accessory sphene.

**Groundmass**   The background, generally finer grained material, that constitutes the main bulk of a rock, and within which larger crystals or other objects may reside.

**Growth twin**   Primary (or growth) twins represent twins present in a given crystal that formed at the time of crystal growth.

**Gypsum plate**   The gypsum plate (retardation $\Delta = 550nm$), is one of several different accessory plates that can be inserted into the beam of cross-polarised light above the analyser, as an aid to determining the "fast" or "slow" direction of a mineral or its optic sign from an interference figure. (see 1.4.5 and Fig. 1.23)

**Hornfels**   A hard, fine to medium grained granoblastic rock produced by high grade contact metamorphism.

**Hour-glass structure**   A structure common in chloritoid (Fig. 1.18c) and certain other minerals, consisting of a dense mass of fine grained (usually opaque) inclusions arranged in the form of an 'hour-glass'. It forms due to influence of the host crystal structure. Aegerine-augite crystals also commonly show an hour-glass structure, somewhat like sector-zoning, typically with a more augite-rich core progressing to a more aegerine-rich margin.

**Inclined extinction**   Minerals showing inclined extinction (e.g. clinopyroxenes, clinoamphiboles), have varying levels of angular difference between the optic axes and the crystallographic axes. The extinction angle (maximum) of a mineral is defined as the maximum difference between the γ -axis (optical axis) and the z-axis (crystallographic axis) (see Figs. 1.20-1.22).

**Inclusion**   A solid or fluid phase totally enclosed within a mineral (e.g. fluid inclusions in vein minerals, matrix phases in porphyroblasts (see Image 277), and melt within phenocrysts).

**Interference figure**   In conoscopic illumination, the light passing through a mineral specimen diverges on exit to produce a conical beam. For favourable oriented crystals (with respect to optic axes), this beam produces an interference figure on the surface of the objective lens. The interference figure comprises a series of dark lines ("isogyres"), relating to the optic axes; a simple cross for uniaxial minerals and two curved black lines for biaxial minerals. By insertion of a Bertrand lens, the interference figure is projected to the eye-piece (see 1.4.6 and Figs. 1.25 & 1.26).

**Isogyres**   The broad black bands (curves) marking areas of extinction in interference figures (see Figs. 1.25 & 1.26).

**Isotropic**   Crystallographically and optically homogeneous crystals (i.e. minerals of the cubic system). Such minerals appear black in cross-polarised light (see Fluorite in Images 87-88). This is because the two perpendicular light rays created by the substage polariser pass through isotropic crystals at the same velocity (same refractive index in all directions), and thus cancel when recombined when the analyser is inserted.

**Lamellar twinning**    Repeated twinning of a crystal about a particular crystallographic plane (e.g. (010) in plagioclase), to produce an array of alternately arranged bands within the mineral, recognised by differences in extinction position (see Figs. 1.17b and 1.22).

**Length fast and length slow crystals**    Polarised light hits the underside of the thin-section as a wave vibrating in a single plane. However, immediately on entering the crystal the light splits into two rays (ordinary ray and extraordinary ray), vibrating in separate planes at right-angles to each other. These rays move at different velocities as a function of the crystallographic structure and atomic arrangement. They are generally referred to as the "fast ray" and "slow ray". Length-slow crystals are those which have the length of the crystal parallel to the slow ray, whilst for length fast crystals the reverse is true. Length slow rays always have a higher refractive index than the fast ray. The length "slow" or "fast" character can be determined with the use of accessory plates (mica plate, gypsum plate, quartz wedge) (see 1.4.5 and Figs. 1.23-1.24).

**Marble**    A metamorphosed limestone, with a mineral assembleage dominated by calcite (see Fig. 1.9 and Image 91).

**Metamorphic facies**    A metamorphic facies is a subdivision of metamorphic conditions in P-T space on the basis of diagnostic mineral assemblages which have been shown by experimental and field observations to characterise a specific range of P-T conditions for a particular compositional group of rocks (in association with $H_2O$ fluid).

**Metamorphism**    The mineralogical, chemical, and structural adjustment of solid rocks to physical and chemical conditions which have generally been imposed at depth below the surface zones of weathering and cementation, and which differ from the conditions under which the rocks in question originated (Bates & Jackson, 1980).

**Metapelite**    A metamorphosed pelitic rock (see pelite).

**Metasomatism**    Metamorphism involving modification of the bulk rock chemistry by influx or removal of chemical components via a fluid phase (e.g. widespread potassium metasomatism associated with granite intrusions).

**Mica ($\lambda/4$) plate**    The mica ($\lambda/4$) plate (retardation $\Delta = 150nm$), is one of several different accessory plates that can be inserted into the beam of cross-polarised light above the analyser, as an aid to determining the "fast" or "slow" direction of a mineral, or its optic sign from an interference figure (see 1.4.5 and Fig. 1.23).

**Mica-fish**    'Fish'- or 'lozenge'-shaped mica porphyroclasts/ porphyroblasts aligned within a finer grained schistose matrix. They are characteristic of phyllonites and highly deformed schists, and can be used to determine shear-sense.

**Microperthite**    The microscopic intergrowth of plagioclase and K-feldspar, comprising exsolved fine plagioclase inclusions enclosed within K-feldspar (see Image 116a). Common in high grade metamorphic rocks and plutonic igneous rocks.

**Microstructure**    The geometric arrangement and interrelationships between grains and internal features of grains.

**Migmatite**    A coarse grained heterogeneous rock type characteristically with irregular and discontinuous interleaving of leucocratic granitoid material (leucosome) and residual high-grade metamorphic material (restite). Migmatites are often intensely folded and heavily veined. General opinion considers most migmatites to have developed by *in-situ* anatexis.

**Multiple lamellar twinning**  Multiple lamellar (or polysynthetic) twinning forms by repeat alternations of the crystal lattice about a particular plane in the crystal lattice (e.g. (010) in plagioclase). The regularity is such that alternate lamellae are identically oriented and thus have the same interference colour (see Figs. 1.17b, 1.22 and Image 136a).

**Myrmekite**  A symplectic intergrowth of vermicular quartz and plagioclase resulting from the retrograde replacement of K-feldspar (see Image 116b).

**Omphacite**  Sodium-rich variety of the clinopyroxene augite. An essential and characteristic mineral of eclogites (see Image 146a-b).

**Orthoscopic illumination**  The standard arrangement when viewing a thin-section using a petrological microscope. This involves the various light rays moving in parallel paths upwards through the lens system of the microscope (and through the thin-section) up to the eye-piece where an image of the thin-section field of view is projected.

**Pegmatite**  A very coarse grained (often granitic) igneous rock, usually with grain size > 3cm.

**Pelite**  A rock of argillaceous composition with a mineral assemblage dominated by phyllosilicate minerals. Original sedimentary rocks are mudstones and siltstones, which when metamorphosed become slates, phyllites, schists etc.

**Peristerites**  Sodic plagioclases consisting of microscopic and sub-microscopic intergrowths of albite and oligoclase. Such feldspars characterise the greenschist facies - amphibolite facies transition.

**Perthite**  A feldspar intergrowth with plagioclase inclusions enclosed within K-feldspar. Common in high grade metamorphic rocks and plutonic igneous rocks (see Image 116a).

**Phenocryst**  An igneous term used for larger crystals set within a finer grained groundmass (see Fig. 1.4a and Image 143).

**Pinnitisation**  The retrogression, especially of cordierite, to an ultra fine grained green or yellow felty mixture of muscovite and chlorite (modified from Deer et al, 1992) (see Image 299).

**Pleochroism**  The ability of an anisotropic crystal to differentially absorb various wavelengths of transmitted light in various crystallographic directions, and thus show different colours in different directions (see Fig. 1.6). As the microscope stage is rotated in plane-polarised light the mineral shows a colour change, and is said to be pleochroic (modified after Bates & Jackson, 1980).

**Poikilitic**  An igneous term used to describe large crystals (phenocrysts) of one mineral enclosing numerous small crystals of one or more other minerals.

**Poikiloblastic**  A metamorphic term used to describe porphyroblasts with abundant mineral inclusions (see Fig. 1.4c).

**Polygonal structure**  A term for rocks containing crystals (usually equigranular) with polygonal shapes (commonly five or six-sided) and dominantly straight boundaries meeting at triple-points (modified from Spry, 1969). (see Image 279a, and accompanying diagram 6k).

**Polymorph**  One of two or more crystallographic forms of the same chemical substance. For example, there are three common polymorphs of $Al_2SiO_5$ (i.e. andalusite, kyanite and sillimanite).

**Polysynthetic twinning**  (see multiple lamellar twinning)

**Porphyritic**  Igneous rock texture, comprising larger crystals (phenocrysts) in a finer grained groundmass.

**Porphyroblast**    A metamorphic mineral that has grown to a much larger size than minerals of the surrounding matrix (see Fig. 1.10).

**Porphyroblastesis**    The growth of porphyroblasts.

**Porphyroblastic**    A term used to describe a metamorphic rock with large crystals (porphyroblasts) grown within a finer grained matrix.

**Porphyroclast**    A large relict crystal, or crystal fragment in a fine grained matrix of a deformed rock.

**Porphyroclastic**    A term used to describe rocks with abundant porphyroclasts (e.g. mylonites).

**Pressure shadow**    (see Strain shadow)

**Primary twins**    Twins developed during crystal growth (see 1.4.2).

**Pseudomorph**    A mineral or aggregate of minerals having the form of another mineral phase being replaced. A pseudomorph is described as being "after" the mineral whose outward form it has (e.g. chlorite after garnet). Pseudomorphing is a gradual process such that at the arrested stage seen in thin-section the pseudomorph may be 'partial' or 'complete' (see Images 261 & 278).

**Quartz wedge**    One of several different accessory plates that can be inserted into the beam of cross-polarised light above the analyser, as an aid to determining the "fast" or "slow" direction of a mineral, or its optic sign from an interference figure. The quartz wedge is a tapered wedge of a single large quartz crystal, cut and enclosed between two glass plates. It is inserted thin-end first, and produces progressively higher interference colours as thicker and thicker sections of the wedge are introduced.

**Reaction rim**    A monomineralic or polymineralic rim (commonly of hydrous phases) surrounding the core of another phase in the process of retrogression

**Regional (orogenic) metamorphism**    Metamorphism affecting large areas of the Earth's crust and commonly associated with collisional orogeny. Regional metamorphic rocks commonly exhibit complex interrelationships between mineral growth and deformation.

**Relief (of minerals)**    The "relief" of a mineral describes the extent to which it stands out from the background when viewed in thin-section. Relief is directly related to the refractive index of the mineral concerned, and the difference in refractive index compared with surrounding minerals. The higher the refractive index of a mineral, the higher the relief (see 1.3.5, Table 1.2, Figs. 1.7 & 1.10).

**Retrograde metamorphism**    Metamorphic changes in response to decreasing pressure and/or temperature conditions.

**Retrogression**    Modification of the primary mineral assemblage due to waning P-T conditions and/or changing fluid chemistry. This process typically involves partial or complete replacement of high grade largely anhydrous phases by lower grade hydrous phases.

**Saussuritisation**    The retrogressive replacement of anorthitic plagioclase in a fine grained aggregate of epidote group minerals and sericite (± calcite).

**Schist**    Metamorphic rock, typically of pelitic or semi-pelitic composition, with a well developed schistosity (see Images 23, 81, 184, 202b).

**Schistosity**    A planar structure defined by the alignment of inequant minerals such as micas and amphiboles and where individual minerals are discernable in hand specimen. Rocks showing this structure are termed schists. Such a structure is common in regional metamorphic and blueschist facies metapelites and metabasites.

**Secondary twins**    Twins that have formed subsequent to crystal growth. This includes *inversion twins* formed due to change in crystal habit as a result of instability of the initial structure with changing P-T, and *deformation twins* formed in response to deformation of the crystal lattice (see 1.17d).

**Sector twinning (or sector-zoning)**    Geometrically arranged triangular, or wedge-shaped, crystallographically induced twins, characteristic of a small selection of minerals, especially cordierite (Fig. 1.18e). Although the sector-twinning (also termed "sector zoning", "sector-trilling") seen in cordierite of contact metamorphic aureoles was originally interpreted as a growth-twinning feature, it is now widely accepted as resulting from the transformation of metastable high-temperature hexagonal cordierite to stable low-temperature orthorhombic cordierite.

**Sericitisation**    The alteration of a mineral or minerals to an aggregate of fine-grained white mica, known as sericite (see Image 263).

**Serpentinite**    A retrogressed ultramafic rock with a mineral assemblage comprised largely of serpentine minerals.

**Serpentinisation**    A process involving the conversion of high temperature minerals (especially olivine) to an aggregate of serpentine (see Image 298). This is common in retrogressed ultramafic rocks, and takes place at temperatures below 500°C and often <300°C in the presence of aqueous fluids.

**Simple twin**    A mirroring of the crystal structure each side of a particular crystallographic plane as the crystal has grown. Because the two halves of the twin are differently oriented within the plane of the thin-section they are easily recognised in cross-polarised light due to different interference colours (see Fig. 1.17a).

**Skarn**    A rock formed during contact metamorphism or regional metasomatism by reaction between carbonate rocks and fluids rich in elements such as iron and silica. Skarns have an assemblage dominated by calc-silicate minerals often in association with magnetite (see Image 325).

**Skeletal crystals**    Individual crystals that have nucleated and grown between the grain boundaries of other crystals (particularly quartz) to form a skeletal or mesh-like network of thin interconnected strands (see Image 320).

**Spherulitic**    A term describing a sub-spherical mass of acicular crystals radiating from a common point (see Image 308).

**Steatisation**    A process involving the retrogression of ultrabasic rocks to an assemblage comprised largely of talc.

**Straight extinction**    For minerals with straight (=parallel) extinction, there is parallelism between the crystallographic axes and optic axes. The extinction angle of a crystal is obtained (in XPL) by measuring the angular difference between the long-axis of the crystal when aligned with N-S cross-wires, and the position where the crystal goes into extinction. In the case of straight extinction the angle is 0°, i.e. extinction is parallel with the N-S cross-wires (see Fig. 1.19).

**Strain shadow**    A region of low strain immediately adjacent to a porphyroblast. It is located perpendicular to the principal axis (axes) of compression, and is largely protected from deformation by the higher rigidity (competence) of the porphyroblast (e.g. garnet, pyrite) relative to the matrix (e.g. quartz and phyllosilicates).

The porphyroblast and matrix may become detached along their contact, the space being filled with new crystals (e.g quartz) showing fibrous growth forms in the direction of extension (if not recrystallized at higher temperatures). (see Image 270)

**Sub-grains**   Sub-grains develop in deformed crystals, and are areas misoriented by a few degrees relative to the parent grain. They are picked out by standard optical microscopy since they pass into extinction at a slightly different position to the parent grain, although lacking sharply defined boundaries. Sub-grains are strain free, and form during recovery.

**Subhedral**   A term used to describe metamorphic crystals (especially porphyroblasts) with moderately good form, and some well-formed crystal faces. (see Fig. 1.3c)

**Symplectite**   Complex and intimate intergrowth of two or sometimes three simultaneously co-nucleating phases, one of which is usually vermicular or rodded in form. Such structures are especially common in retrogressed granulites.

**Twin**   A polycrystalline unit composed of two or more homogenous portions of the same crystal species mutually oriented according to certain simple laws (see Fig. 1.17).

**Twinkling**   By virtue of pronounced differences in refractive indices in different crystallographic orientations, some minerals, most notably carbonate minerals (e.g. calcite and dolomite) show marked change in relief when the microscope stage is rotated. This produces an effect known as "twinkling", where the relief of individual crystals rises and falls as the stage is turned (see Fig. 1.9).

**Ultramafic rock**   Igneous rocks dominated by ferromagnesium minerals (e.g. olivines, pyroxenes) (see Image 306).

**Undulose extinction**   A term describing a property of deformed crystals (especially quartz) in which poorly defined zones of extinction are seen to sweep across individual crystals when the microscope stage is rotated. This feature is caused by deformation/distortion of the crystal lattice.

**Uralitisation**   A process of alteration involving the pseudomorphic replacement of primary igneous pyroxenes by secondary metamorphic/hydrothermal amphiboles (often tremolite, actinolite or hornblende).

**Vein**   A fracture or microfracture filled with crystalline material (see Fig. 1.12).

**Vesicle**   Rounded, ellipsoidal or irregular small holes in lavas and near surface volcanic rocks, where volatiles have been exsolved from the magma during cooling and decompression (see Image 265). If filled with a secondary mineral they are termed amygdales.

**Uniaxial interference figure**   In conoscopic illumination, the light passing through a mineral specimen diverges on exit to produce a conical beam. In uniaxial crystals (e.g. quartz) cut perpendicular to the z- crystallographic axis ($\equiv$ c-axis), this beam produces an interference figure on the surface of the objective lens. By insertion of a Bertrand lens, the interference figure is projected to the eye-piece. The interference figure comprises a series of dark lines ("isogyres"), relating to the optic axes, and for uniaxial minerals produces a simple cross (see 1.4.6 and Fig. 1.25).

**Vermicular**   A term for crystals with worm-like form (see Image 116b).

# Appendix 4: Mineral abbreviations

The mineral abbreviations used throughout the present key are those given in Tables 2.12.1 and 2.12.2 of Fettes & Desmons (2007). In broad terms these abbreviations follow the widely accepted list of Kretz (1983), but with updates and revision. The fully revised list of Fettes & Desmons (2007) is that recommended by the Subcommission on the Systematics of Metamorphic Rocks (SCMR), and a partial version of this list is given below. The abbreviations used by Bucher & Frey (1994) and Barker (1998) have a similar basis and show broad agreement.

Ab = albite
Act = actinolite
Aeg = aegirine
Afs = alkali feldspar
Agt = aegirine-augite
Alm = almandine
Aln = allanite
Als = aluminosilicate
An = anorthite
And = andalusite
Anh = anhydrite
Ank = ankerite
Anl = analcite (analcime)
Ann = annite
Ap = apatite
Arf = arfvedsonite
Arg = aragonite
Atg = antigorite
Ath = anthophyllite
Aug = augite
Brc = brucite
Brl = beryl
Brt = barite (barytes)
Bt = biotite
Cal = calcite
Cbz = chabazite
Ccn = cancrinite
Chl = chlorite
Chn = chondrodite

Chr = chromite
Chu = clinohumite
Cld = chloritoid
Cls = celestine
Coe = coesite
Cph = carpholite
Cpx = clinopyroxene
Crd = cordierite
Crn = corundum
Crs = cristobalite
Cst = cassiterite
Ctl = chrysotile
Cum = cummingtonite
Czo = clinozoisite
Dee = deerite
Di = diopside
Dol = dolomite
Drv = dravite
Dsp = diaspore
Elb = elbaite
En = enstatite
Ep = epidote
Fa = fayalite
Fl = fluorite
Fo = forsterite
Fs = ferrosilite
Fsp = feldspar
Gbs = gibbsite
Ged = gedrite

Gln = glaucophane
Glt = glauconite
Gn = galena
Gp = gypsum
Gr = graphite
Grs = grossular
Grt = garnet
Hbl = hornblende
Hc = hercynite
Hd = hedenbergite
Hem = hematite (haematite)
Hu = humite
Hul = heulandite
Hyn = haüyne
Ill = illite
Ilm = ilmenite
Jd = jadeite
Kfs = K-feldspar
Kln = kaolinite
Krn = kornerupine
Krs = kaersutite
Ky = kyanite
Laz = lazulite (lazurite)
Lct = leucite
Lm = limonite
Lmt = laumontite
Lpd = lepidolite
Lws = lawsonite
Lz = lizardite
Mag = magnetite
Mar = mariolite
Mc = microcline
Mei = meionite
Mel = melilite
Mgs = magnesite
Mnt = montmorillonite
Mnz = monazite
Mrg = margarite
Ms = muscovite
Mtc = monticellite
Mul = mullite
Ne = nepheline
Nsn = nosean
Ol = olivine
Omp = omphacite
Op = opaque mineral
Opx = orthopyroxene
Or = orthoclase
Osu = osumilite

Per = periclase
Pg = paragonite
Pgt = pigeonite
Phg = phengite
Phl = phlogopite
Pl = plagioclase
Pmp = pumpellyite
Prh = prehnite
Prl = pyrophyllite
Prp = pyrope
Prv = perovskite
Px = pyroxene
Py = pyrite
Qtz = quartz
Rbk = riebeckite
Rds = rhodochrosite
Rt = rutile
Sa = sanidine
Scp = scapolite
Sd = siderite
Sdl = sodalite
Ser = sericite
Sil = sillimanite
Sme = smectite
Sp = sphalerite
Spd = spodumene
Spl = spinel
Spr = sapphirine
Sps = spessartine
Srl = schorl (schorlite)
Srp = serpentine
St = staurolite
Stb = stilbite
Stp = stilpnomelane
Tlc = talc
Toz = topaz
Tr = tremolite
Trd = tridymite
Ttn = titanite (sphene)
Tur = tourmaline
Uvt = uvarovite
Ves = vesuvianite (idocrase)
Wo = wollastonite
Wrk = wairakiite
Zeo = zeolite
Zo = zoisite
Zrn = zircon
Zwd = zinnwaldite

# References

Adams, A.E., & MacKenzie, W.S. (2001). *A Colour Atlas of Carbonate Sediments and Rocks Under the Microscope*. London, UK, Manson Publishing.

Adams, A.E., MacKenzie, W.S., & Guilford, C. (1984). *Atlas of Sedimentary Rocks Under the Microscope*. Harlow, England, UK, Longman Group Ltd.

Barker, A.J. (1998). *Introduction to Metamorphic Textures and Microstructures*, 2nd ed. Cheltenham, UK, Stanley Thornes.

Bates, R.L., & Jackson, J.A. (1980). *Glossary of Geology*. Falls Church, VA, USA, American Geological Institute.

Bowles, J.F.W., Howie, R.A., Vaughan, D.J., & Zussman, J. (1996). *Rock-Forming Minerals, Volume 5A: Non-Silicates: Oxides, Hydroxides and Sulphides*. London, UK, The Geological Society.

Bucher, K., & Frey, M. (1994). *Petrogenesis of Metamorphic Rocks*. Berlin, Germany, Springer.

Chang, L.L.Y., Howie, R.A., & Zussman, J. (1996). *Rock-Forming Mineral, Volume 5B: Non-Silicates: Sulphates, Carbonates, Phosphates and Halides*. London, UK, The Geological Society.

Clapham, A.R., Tutin, T.G., & Warburg, E.F. (1981). *Excursion Flora of the British Isles*, 3rd ed. Cambridge, UK, Cambridge University Press.

Craig, J.R,. & Vaughan, D.J. (1994). *Ore Microscopy and Ore Petrography*, 2nd ed. New York, NY, John Wiley & Sons.

Deer, W.A., Howie, R.A., Wise, W.S., & Zussman, J. (2011). *Rock-Forming Minerals, Volume 4B: Framework Silicates: Silica Minerals, Feldspathoids and the Zeolites*, 2nd ed. London, UK, The Geological Society.

Deer, W.A., Howie, R.A., & Zussman, J. (1982). *Rock-Forming Minerals, Volume 1A: Orthosilicates*, 2nd ed. London, UK, The Geological Society.

Deer, W.A., Howie, R.A., & Zussman, J. (1986). *Rock-Forming Minerals, Volume 1B: Disilicates and Ring Silicates*, 2nd ed. London, UK, The Geological Society.

Deer, W.A., Howie, R.A., & Zussman, J. (1992). *An Introduction to the Rock-Forming Minerals*, 2nd ed. London, UK, Longman Group.

Deer, W.A., Howie, R.A., & Zussman, J. (2013). *An Introduction to the Rock-Forming Minerals*, 3rd ed. London, UK, The Mineralogical Society.

Deer, W.A., Howie, R.A., & Zussman, J. (1997a). *Rock-Forming Minerals, Volume 2A: Single-Chain Silicates*, 2nd ed. London, UK, The Geological Society.

Deer, W.A., Howie, R.A., & Zussman, J. (1997b). *Rock-Forming Minerals, Volume 2B: Double-Chain Silicates*, 2nd ed. London, UK, The Geological Society.

Deer, W.A., Howie, R.A., & Zussman, J. (2001). *Rock-Forming Minerals, Volume 3: Layered Silicates: Excluding Micas and Clay Minerals*, 2nd ed. London, UK, The Geological Society.

Deer, W.A., Howie, R.A., & Zussman, J. (2004). *Rock-Forming Minerals, Volume 4A: Framework Silicates: Feldspars*, 2nd ed. London, UK, The Geological Society.

Demange, M. (2012). *Mineralogy for Petrologists: Optics, Chemistry and Occurrences of Rock-Forming Minerals*. London, UK, Taylor & Francis Group.

Ehlers, E.G. (1987a). *Optical Mineralogy (Volume 1): Theory and Techniques*. Palo Alto, CA, Blackwell Scientific Publications.

Ehlers, E.G. (1987b). *Optical Mineralogy (Volume 2): Mineral Descriptions*. Palo Alto, CA, Blackwell Scientific Publications.

Fettes, D., & Desmons, J., eds. (2007). *Metamorphic Rocks: A Classification and Glossary of Terms*. Cambridge, UK, Cambridge University Press.

Fleet, M.E. (2009). *Rock-Forming Minerals, Volume 3A: Micas*. London, UK, The Geological Society.

Ixer, R. (1990). *Atlas of Opaque and Ore Minerals in Their Associations*. Milton Keynes, UK, Open University Press.

Johannsen, A. (1908). *A Key for the Determination of Rock-forming Minerals in Thin-Sections*. New York, NY (Kessinger Publishing edition), John Wiley & Sons.

Kerr, P.F. (1977). *Optical Mineralogy*, 4th ed. New York, NY, McGraw-Hill, pp. 492.

Kretz, R. (1983). Symbols for rock-forming minerals. *American Mineralogist*, 68, 277–279.

MacKenzie, W.S., & Adams, A.E. (1994). *A Colour Atlas of Rocks and Minerals in Thin-Section*. London, UK, Manson Publishing Ltd.

MacKenzie, W.S., Donaldson, C.H., & Guilford, C. (1982). *Atlas of Igneous Rocks and Their Textures*. Harlow, England, UK, Longman Group Ltd.

MacKenzie, W.S., & Guilford, C. (1980). *Atlas of Rock-Forming Minerals in Thin-Section*. Harlow, England, UK, Longman Group Ltd.

Mange, M.A., & Maurer, H.F.W. (1992). *Heavy Minerals in Colour*. London, UK, Chapman & Hall.

Shelley, D. (1993). *Igneous and Metamorphic Rocks Under the Microscope*. London, UK, Chapman & Hall.

Spry, A. (1969). *Metamorphic Textures*. Oxford, UK, Pergamon Press.

Stace, C. (1999). *Field Flora of the British Isles*. Cambridge, UK, Cambridge University Press.

Vernon, R. (1976). *Metamorphic Processes*. London, UK, George Allen & Unwin.

Vernon, R.H. (2004). *A Practical Guide to Rock Microstructure*. Cambridge, UK, Cambridge University Press.

Wenk, H-R., & Bulakh, A. (2004). *Minerals Their Constitution and Origin*. Cambridge, UK, Cambridge University Press.

Yardley, B.W.D., MacKenzie, W.S., & Guilford, C. (1990). *Atlas of Metamorphic Rocks and Their Textures*. Harlow, England, UK, Longman Group Ltd.

# Index

Actinolite *10, 23*, **1** (39), **2** (61, 63)
Adularia **2** (57)
Aegirine **1** (48), **2** (59)
Aegirine-Augite **1** (48), **2** (62)
Aenigmatite **1** (40), **2** (68), **7** (140, 141)
Albite **2** (54, 55), **3** (75), **6** (116, 127)
Allanite (*Orthite*) **4** (95), **5** (101)
Almandine **6** (112), **7** (140)
Analcite **1** (44), **6** (120)
Andalusite *17, 20–21, 25–27*, **1** (46), **3** (72–73, 79, 81), **5** (102), **6** (122, 124, 128)
Andesine **2** (55)
Andradite **6** (112)
Anhydrite *17*, **1** (51), **3** (85)
Ankerite **1** (42)
Anorthite **2** (54–55)
Anorthoclase **1** (45)
Anthophyllite **1** (37), **3** (76)
Antigorite **3** (78), **7** (138)
Apatite *8*, **6** (120)
Aragonite **5** (98), **6** (112)
Arfvedsonite **2** (59)
Astrophyllite **4** (95)
Augite *16, 17, 18, 22–23*, **1** (50), **2** (62, 70)
Axinite **5** (103)

*Barkevikite* (see Kaersutite)
Barytes **1** (43), **3** (73, 74, 82)
Beryl **6** (120, 121)
Biotite *6, 8, 10, 11, 13, 16, 22*, **4** (93, 94), **6** (118), **7** (144)
Brucite **3** (81)
Bytownite *23*, **2** (54–55)

Calcite *12, 13, 14, 19–20*, **1** (42), **5** (98), **6** (112, 129)
Cancrinite *17*, **3** (83)
Cassiterite *17*, **5** (106–107), **7** (137)
Celestine **1** (51), **3** (82)
Chabazite **5** (109)
Chalcedony **6** (125)
*Chiastolite* (see Andalusite)
Chlorite *16, 18*, **4** (93), **7** (137, 138, 143)

Chloritoid *20–21, 27*, **1** (39), **4** (92)
Chondrodite **2** (61), **5** (100, 108), **7** (135)
Chromite *8* (152)
Chrysotile **6** (130)
Clay minerals **3** (79, 81), **6** (122, 128), **7** (134)
Clinohumite **5** (100, 108), **7** (135)
Clinopyroxene **2** (66)
Clinozoisite **3** (75), **6** (116, 127)
Cordierite *20–21*, **6** (126, 127)
Corundum **1** (43), **5** (104), **6** (114, 115, 117)
Crossite **1** (38), **2** (60)
Cummingtonite **1** (37), **2** (65)

Deerite **7** (139)
Diaspore **3** (77), **4** (88)
Diopside *19, 20–21*, **1** (46, 47), **2** (67)
Dolomite *12, 14, 17, 20*, **1** (42), **5** (98), **6** (112)
Dravite (see Tourmaline)
Dumortierite **4** (91), **7** (133)

Enstatite **1** (46), **3** (73)
Epidote **4** (90), **6** (127), **7** (135)
Eudialyte **6** (120, 121)

Fayalite **5** (99)
Fe-Calcite **1** (42)
Ferrosilite **4** (89)
*Fibrolite* (see Sillimanite)
Forsterite (see Olivine)
Fluorite *5*, **1** (36, 41)
Fuchsite **4** (93)

Galena *8* (149)
Garnet *8, 11, 13*, **5** (99), **6** (112, 128), **7** (140)
Gibbsite **2** (65), **6** (122)
Glass (hole in slide) **6** (124)
Glauconite **7** (137)
Glaucophane **1** (38), **2** (60), **3** (75), **4** (91), **7** (133)
Goethite **7** (138, 143)
Graphite *8* (152)
Grossular **5** (102), **6** (112, 114)

Grunerite **1** (37), **2** (64)
Gypsum **3** (80), **6** (129)

Haematite **7** (138, 139, 143), **8** (148)
Haüyne **1** (44), **5** (108)
Hedenbergite **1** (50), **2** (69)
Hemimorphite **3** (82)
Hercynite **7** (136)
Heulandite **3** (81)
Hole in thin-section (glass) **6** (124)
Hornblende *8, 9,* **1** (39, 40), **2** (61, 62, 63, 68)
Humite **5** (100, 108), **7** (135)
Hypersthene **1** (49), **4** (89)

*Iddingsite* **7** (143)
*Idocrase* (see Vesuvianite)
Illite **3** (79), **6** (122, 123)
Ilmenite **8** (147, 149)
Ilmenomagnetite **8** (149)

Jadeite **1** (47), **2** (66)

K-feldspar *20*
Kaersutite **1** (40), **2** (62, 63, 68)
Kaolinite **6** (128)
Kyanite **1** (43), **2** (58, 64), **3** (75), **4** (88)

Labradorite *23,* **2** (54, 55)
Laumontite **6** (122, 130)
Lawsonite **1** (41), **3** (75)
Lazurite **7** (132)
Lepidocrocite **7** (138)
Lepidolite **3** (84, 85)
Leucite *17,* **6** (119, 126)
Limonite **7** (138)

Magnesite **1** (42)
Magnetite **2** (62), **8** (151)
Margarite **5** (104)
Melanite **7** (140)
Melilite *16,* **5** (102, 109)
Mesolite **3** (78), **6** (119)
Micrite **6** (112)
Microcline *12, 19–20,* **1** (45)
Microperthite **2** (54)
Monazite **2** (58), **3** (77), **5** (105), **6** (118)
Monticellite **5** (104)
Muscovite *8, 11, 13, 16, 17, 30,* **3** (84),
    **6** (123)
Myrmekite *6,* **2** (54)

Natrolite **3** (78, 82), **6** (119, 122)
Nepheline **3** (80), **5** (107), **6** (126)
Nosean **1** (44), **5** (108)

Oligoclase **2** (55)
Olivine (Forsterite) *8, 17,* **5** (105), **6** (118),
    **7** (141)

Omphacite **1** (47, 49), **2** (69, 70)
*Orthite* (see Allanite)
Orthoclase **2** (57)
Orthopyroxene **4** (89)
*Ottrelite* (see Chloritoid)
Oxo-hornblende *7,* **1** (40), **2** (62, 63)

Palagonite **7** (134, 144)
Paragonite **3** (84), **6** (123)
Periclase **1** (44)
Perovskite **7** (140, 141)
Perthite **2** (54)
Phengite **2** (70), **4** (93)
Phlogopite *16, 17, 18, 30,* **4** (94)
Piemontite *10,* **4** (89), **7** (132)
Pigeonite **1** (49, 50), **2** (66, 67, 69, 70)
*Pinite* **7** (138)
Pistacite (see Epidote)
Plagioclase *6, 19–20, 23–24,* **2** (54–55),
    **6** (127)
Pleonaste **7** (136)
Prehnite *15,* **3** (76, 82)
Pumpellyite **2** (61, 63), **7** (136)
Pyrite *5,* **6** (125), **8** (150)
Pyrope **6** (112)
Pyrophyllite **3** (84), **6** (123)

Quartz *8, 11, 13, 29,* **2** (54), **6** (125, 127,
    128, 129)

Rhodochrosite *17,* **1** (43), **5** (98)
Riebeckite **1** (38), **2** (60), **4** (91)
Rutile *16,* **1** (49), **4** (88), **7** (141, 143, 144)

Sanidine **2** (57), **3** (79)
Sapphirine **7** (133)
Scapolite **3** (83)
Schorl (see Tourmaline)
Schorlomite **7** (140)
Scolectite **2** (63)
Sericite **3** (79, 84), **6** (123, 128), **7** (138)
Serpentine **3** (78), **6** (130), **7** (137)
Siderite **1** (42), **7** (142)
Sillimanite *11, 16, 18,* **5** (102, 103),
    **6** (115, 117)
Smectite **3** (81), **6** (122), **7** (134, 143, 144)
Sodalite **1** (44), **5** (108), **7** (132)
Spessartine *6,* **5** (106), **6** (112), **7** (140)
Sphalerite **1** (36), **5** (106)
*Sphene* (see Titanite)
Spinel **7** (136)
Spodumene *6, 13,* **1** (36, 47), **2** (66)
Staurolite *5, 7, 27,* **4** (90), **5** (99), **7** (134)
Stilbite **3** (81)
Stilpnomelane *22,* **4** (94)

Talc *17,* **3** (83)
Thomsonite **3** (78), **6** (119, 125)

Titanaugite *7, 20–21*, **1** (48), **2** (68)
Titanite *13*, **5** (106), **7** (142)
Topaz **3** (72, 80), **6** (121)
Tourmaline (Dravite) **5** (100), **7** (135)
Tourmaline (Schorl) *6, 8, 9, 22, 27, 29,*
    **4** (92), **5** (101), **7** (136)
Tremolite **1** (37), **2** (65)

Uvarovite **6** (112)

Vesicle **6** (124)
Vesuvianite *6, 16, 20–21*, **5** (102), **6** (114)
Volcanic glass **6** (124)

Wollastonite **3** (74), **6** (116, 122)

Yoderite **5** (100)

Zeolite *17*, **2** (63), **3** (78, 80), **6** (119, 122,
    125, 130)
Zinnwaldite **3** (85)
Zircon **6** (118), **7** (144)
Zoisite *16, 18*, **3** (72), **6** (113)